TURNING POINTS

TURNING POINTS

How Critical Events Have Driven Human
EVOLUTION, **LIFE**, and **DEVELOPMENT**

WITHDRAWN

KOSTAS KAMPOURAKIS

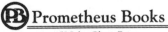 Prometheus Books

59 John Glenn Drive
Amherst, New York 14228

Published 2018 by Prometheus Books

Cover images © Media Bakery
Cover design by Liz Mills
Cover design © Prometheus Books

Inquiries should be addressed to
Prometheus Books
59 John Glenn Drive
Amherst, New York 14228
VOICE: 716–691–0133 • FAX: 716–691–0137
WWW.PROMETHEUSBOOKS.COM

22 21 20 19 18   5 4 3 2 1

Library of Congress Cataloging-in-Publication Data Pending

Printed in the United States of America

To Katerina; meeting her was contingent *per se*,
and everything has been contingent *upon* it ever since.

CONTENTS

CONTENTS

PREFACE

My own life, like everyone else's, has been full of critical events that have driven it in the direction it has taken. As a child, I was prone to believe that there is purpose in one's life. A main reason for this belief was that my father, a Greek army officer who happened to serve in the Cypriot armed forces, was severely injured—and almost died—during the Turkish invasion to Cyprus in July 1974, while my mother was pregnant with me. He survived because he was lucky enough to be found by a comrade and be transported to the hospital, where he was, again, lucky enough to avoid a leg amputation by mistake—events totally contingent. He is now seventy-two years old and in full health, and he still believes that there was a purpose in all that. I am his lucky child, he often says, as he believes that he did not die there for a purpose: in order for me not to grow up as an orphan. His beliefs and my own life were certainly contingent upon the events of July 1974. My life would have been totally different had he died there. To say the least, my young brother would not have been born. For these reasons, I initially accepted that there must have been a purpose for his survival: that I would not grow up as an orphan and that my brother would be born. This was what my father believed; but I must note that he never tried to impose any kind of fatalistic thinking on my brother and me. It was just what he believed.

But as soon as I started my undergraduate studies in biology, my beliefs and worldviews changed—radically. I soon developed a different understanding of life, one dominated by contingencies and conjunctures, and I stopped believing in a plan and purpose in life. That was an important shift. Being a city boy, I had only a limited interaction with domestic animals, when we visited my father's parents in their village in Crete during summer. Like many others, I also grew up watching all of those anthropomorphic, happy-animal, happy-ending children's films. But as an undergraduate student, I started a more sophisticated exploration of nature, not as an amateur naturalist but through books and documentary films about nature. The extreme violence that existed in nature made me reconsider my views about the principles that govern life, with death being a major one. An event of this kind also dominated my childhood. Two years after I was born, my father's brother died from a heart attack at the age of twenty-nine. Everyone was talking about him and about how unlucky he was. As a child, I always found it hard to understand why he died; why I never had the chance to get to know that person who was admired by everyone who knew him and who had just gotten married a few weeks before his sudden death. I wanted to hope that there was some good reason for this, but, eventually, I realized that there was not; he just died (needless to say, I doubt that my father ever saw any purpose in that devastating event like he saw in his surviving his injuries). Any remaining thoughts I had about plan and purpose in nature had vanished.

My whole professional life, and my writing hobbies as well, have to do with understanding, teaching, and learning biology. One could have thought that I was meant to do that, because I do practically nothing else, except for spending time with my family and friends. Reading and writing about biology is a defining feature of who I am. However, how I ended up studying biology is a funny example of the impact of contingencies on one's life. I finished school before I was eighteen years old, and participated in the Greek national exams in 1992. At that time, university candidates were supposed to be examined on four subjects, and could participate in the exams up to three consecutive times. An important detail: one could submit for some of these subjects the high grades one had gotten in past exams, and then, therefore, be examined only in the remaining subjects. This means that those who were finishing school and were participating in the exams for the first time, being examined in four subjects, had to compete with other candidates who were participating for the second or the third time and were being examined in, for example, only two subjects after having spent a whole year working on them. To show the impact of these conditions one need only look at my peers: out of the ninety first-year undergraduate students of my cohort, I can think of only myself and two or three others who had just finished school at the time. All of the other students were one or two years older than us.

In line with the trend of the time for any reasonably good student interested in the life sciences, my "fate" was to become a medical doctor; this was also the hope and wish of my whole family. I was always interested in biology, but at the time there was not much going on in Greece that was widely known, and so I decided to try to enroll in the study of medicine and see after that what I would do. However, there were already employment issues because the field was inundated with medical doctors (as a result of the I-want-to-become-a-doctor trend); so, following my father's advice, I decided to become a medical doctor in the Greek army. Students in Greece apply not directly to universities but to the Ministry of Education, submitting a file with their preferred university departments—at that time this was done *before* the exams. Therefore, I put the military school for becoming a medical doctor first, the department of biology in Athens second, the department of biology in Thessaloniki third, and the remaining biology departments after that. I also added a few technical schools just to have them. I should note that these departments did not ask for a minimum score for accepting students. Rather, they accepted a certain number of students, for instance, the ninety applicants with the best scores among all of those who had mentioned that department in their application—whatever that score was.

I actually wanted to study biology, but my family and family friends had convinced me that professionally I would be a failure. It so happened that I ended up having a score of 5695 points out of 6400. The last student to enter the military school for medical doctors had 5720 points, so I was out; the last student to enter the biology department in Athens had 5694, one point less than me, even though there were two others between us; and the last student to enter the biology department in Thessaloniki had a slightly lower score. Therefore, I ended up in the department of biology in Athens, where I was living, and after a year as an undergraduate (and a second, failed attempt, because of my limited willingness and motivation, to enter the medical school for army doctors), I decided that I would stay in biology. If I had ended up in the biology department in Thessaloniki, it's unlikely that I would have gone there, because it was not possible at the time for me to study in another city; therefore, I would have probably studied carefully to participate in the national exams the next year. Since I wouldn't have enrolled in university studies, I would have probably been highly willing and motivated and thus would have a higher probability to enter the medical school. Luckily, for medicine and for me, this never happened.

My story shows the combined impact of different kinds of factors in human life. On one hand was the surrounding environment (i.e., how exams for entering the university worked), which would be the equivalent for natural conditions for development and evolution. On the other hand, there were my personal decisions, for instance, to decide to start my studies in biology, and eventually continue and get a biology degree— even though I was not eventually so much satisfied by those studies. Unfortunately, universities are not always the motivating, inspirational, thought-provoking environment one wishes them to be. But the important point is that my decisions to do undergraduate studies in biology made a difference. It was unpredictable, as nobody could have known in advance in which of the university departments that I had mentioned in my application I would end up being accepted. This depended not only on my own score but also on the scores of all the other candidates; I happened to be among the ninety who were qualified for the department of biology in Athens, and I was accepted. Once this happened, it thereafter affected my professional life.

Interestingly, not only my professional life, including writing this book, but also my personal life was influenced by my decision to study biology. It was because of my connection to the department of biology in

Athens that I met my wife, best friend, and companion in life, Katerina, who also spent some time in that department even though she did most of her studies in Paris. Whereas I had already completed my undergraduate studies in biology at that point, Katerina was completing her own undergraduate studies in Athens. After several years in Paris, she decided to come back home and complete her studies there in order to figure out what she would do next. Thus, she was attending classes at the department of biology in Athens, whereas I was not as I had already graduated. Meeting her was an event totally unpredictable, as we met through a person who had also studied at the same department and who was going there regularly at the time, whom I had only met twice before (another contingency: I came to know him because he happened to do his military service under my father), and to whom Katerina happened to talk for the first time on the same day that she also met me (because he and I had already planned to meet and talk shop and she wanted to learn more about job opportunities in Greece)! Katerina and I have been together ever since. This has definitely been the most critical event of my life since then, as it has influenced all the subsequent ones—including, above all, the birth of our children, Mirka and Giorgos. Therefore, I dedicate this book to Katerina for being the center of the most important turning point of my life, which has influenced all the subsequent ones that we have lived through together.

ACKNOWLEDGMENTS

First of all, I want to thank Steven L. Mitchell and Prometheus Books for giving me the opportunity to write this book. This book would not have existed without them. Our collaboration has been wonderful, and I am still impressed by how timely and efficiently they do everything. I am especially grateful to Hanna Etu, Mark Hall, Jill Maxick, Liz Mills, Bruce Carle, Cheryl Quimba, Cate Roberts-Abel, and Jade Zora Scibilia, for all their work until the publication of the present book. Even though I am sure that all of these people did a wonderful job, I owe special thanks to Jade Zora Scibilia, who meticulously copyedited the text. I do look forward to working on a next book with them.

My career as a biology teacher and educator has focused on helping students at all levels, from kindergarten to teacher training programs, acquire a rational and authentic view of life. My students, and their conceptions, have inspired and motivated me to write books, hopefully rigorous and accessible ones, which would help any interested reader understand biological phenomena and the related concepts. It is certainly no coincidence that my two previous books are titled *Understanding Evolution* and *Making Sense of Genes*; unavoidably, the present book shares some of the insights of those two books. Therefore, I am grateful to my former students for being a source of inspiration and motivation. I hope that the present book, which is intended for any thoughtful lay persons interested in biology, will sufficiently explain basic phenomena and share with them my own understanding of life.

I am indebted to John Beatty for his work that has been an inspiration for this book. It was back in 2007, when I had just finished my doctoral dissertation on evolution education, that, while we did not know each other, I just contacted him and proposed to meet him and talk. His landmark paper, "Replaying Life's Tape," had recently been published and I had so much I wanted to discuss with him. I thus visited Vancouver, after a conference at Calgary, and John—one of the nicest people you can ever meet—was not only a great host but also an amazing discussant. It was then that I mentioned to him for the first time the idea about writing a book on contingency. His initial encouragement made me determined to write it. And I am happy that you are now holding it in your hands.

I am also indebted to Jonathan Losos, who graciously sent me a copy of his magnificent book *Improbable Destinies*, which was approaching publication as I was finishing writing the present book; as well as to Bernard Wood, who kindly provided me with figure 16.1 and useful essays related to that. Finally, I am very grateful to Francisco Ayala, Henry Gee, Alan Love, Kevin McCain, Alessandro Minelli, Ronald Numbers, Gregory Radick, Michael Ruse, and Elliott Sober for their comments and suggestions on the manuscript's earlier versions.

Last but not least, I am always grateful to my family, Katerina, Mirka, Giorgos, who bear with me while I am reading or writing or discussing my new book project at home. Writing will always be a hobby for me, even if I ever end up doing it as my main job. Thus, whereas other fathers and husbands deal with the garden or watch sports programs on TV while at home, I am reading and writing for hobby—it is not that bad, is it? Therefore, my wife and children do not miss me, but there must certainly be some other things we could have done together during those book-devoted hours. Lost hours are lost forever, but I hope that one day they may read this book and feel that it was worth the time and effort.

CRITICAL EVENTS AND HISTORICAL OUTCOMES

O ne may have good reasons to worry that the public, broadly con-strued,[1] holds unscientific views about several aspects of human life. For instance, 42 percent of Americans believe that people are born homosexual, a percentage that has increased significantly since 1977, whereas more than half the people in Canada and Great Britain seem to think the same.[2] It has also been found that approximately one in four people in the United States, Canada, and Great Britain believe that the position of the stars and planets can affect their lives.[3] Finally, polls since 1982 until very recently show that over 40 percent of Americans accept the idea that God has somehow created humans in their present form.[4] There is also academic research suggesting that people tend to perceive evolution as a purposeful process, as well as believe that genes determine our traits and disease.[5] These, and other, findings support the conclu-sion that unscientific notions are rather widespread, although one should always be cautious about the possible interpretations of research findings about public opinion.[6]

What is common in all these views? One common aspect is the idea of determinism. People think that genes, or something else innate, deter-mine traits, disease, and even behaviors (such as homosexuality), whereas we know that in most cases these are the outcome of complex interaction between genes and environment; people think that the stars determine our lives, even though the stars are too far to have any empirically measurable effect on us; and people think that God determines how we are and look, even though we exhibit so many features that would make any designer feel embarrassed. Another common aspect of these views is the idea that in all of these cases there is a goal, which could be the outcome of purpose, intention, or design. When people think that genes, stars, or God determine our features and/or aspects of our lives, the underlying assumption is that there is some underlying plan toward some specified end point. However, a close look at the conclusions of research in developmental biology, human history, and evolutionary biology show that the course and outcomes of life are not predetermined based on some kind of plan, but rather that they can be influenced by particular combinations of critical events.

Let me clarify the terms I am using throughout the book. There are three closely related concepts (the first two are often considered synonyms), but they are distinct: *fate*, *destiny*, and *design*. Historically, *fate* and *destiny* have been defined in a variety of ways. However, fate is a concept that denotes that there are several aspects of our lives that we do not choose and do not control. We did not choose our parents or our biological characteristics, for instance. There are indeed some features that we are predisposed to have, such as having two eyes and two ears, or two legs and two arms. We do not anticipate human development to result in wings or beaks. It is only in this sense that a kind of developmental fate is conceivable. However, the impact of fate in our biology is sometimes exaggerated in the minds of people, resulting in the conception that I describe as *genetic fatalism*.[7] This view is explored in chapter 2, and part 2 of the present book provides concrete examples that challenge it.

Destiny is a concept that denotes that one can foresee future outcomes by evaluating particular elements that are already present or that have been present in the past. Based on these, one can make a projection to the future and imagine future outcomes, by envisioning what one could become based on what one already is. For instance, one may predict that a child who is extremely intelligent and has a great interest for nature is destined to become a great scientist. Or that someone who has a talent in music or arts is destined to become a musician or an artist. For some people believing in astrology, even the day when one was born is informative for what will happen in the future. The difference with fate is that one's destiny cannot be just realized without effort and decisions. However, the idea of destiny is often misconstrued by people, who do not realize the impact of events within and outside our control and who think as if destiny is a predetermined outcome that will emerge one way or the other. This view is explored in chapter 3, and part 3 provides concrete examples that challenge it.

Finally, design is the idea that the world as we see it is the intentional work of a conscious, intelligent agent (usually God, however conceived) who designed it purposefully. As a result of this design, the physical world has all the properties that are necessary for the emergence of life. A related idea is that the universe also has all of those conditions that are necessary for the emergence of sentient beings such as ourselves, so that our presence in this world was inevitable under these conditions. Some people also claim that organisms exhibit such an enormous complexity in their structures and functions so that the most plausible explanation is that they were specially created and intelligently designed by God. This central idea here is that the complexity of organisms is so enormous that their emergence through natural processes is simply inconceivable; therefore, according to this view, organisms can only have been designed by an intentional and intelligent agent. This view is explored in chapter 4, and part 4 provides concrete examples that challenge it.[8]

In the present book I argue that genetic fatalism, destiny, and intelligent design are insufficient and illegitimate accounts for human development, human life, and human evolution. In all cases, outcomes are better accounted for by considering robust processes (developmental, historical, and evolutionary), as well as critical events that affected which one of several possible directions these processes took. To support this argument, I draw on published research that should nevertheless be considered as a representative and indicative sample of huge bodies of research, rather than an exhaustive account. I focus on studies reporting academic research, and in all cases I am citing research articles that I find indicative of the respective findings and relevant to the argument I am developing in this book. This is especially important to keep in mind both about the research about the conceptions that people hold presented in the first part of the book, and about the research involving organisms, DNA molecules, fossils, and so on presented in the rest of the book.

In all cases, there are particular limitations that relate both to the object of the analysis itself and to the methods used. These limitations usually have an impact on the data obtained, and this should be taken into account in how these data are interpreted. This is especially important for the first part of the book, which presents research on human conceptions. As some researchers nicely put it, most people in the world are not WEIRD: Western, Educated, Industrialized, Rich, and Democratic. Yet this is the kind of people involved in much of the research presented in the first part of the present book (mostly from Europe and North America; the country in which a study took place is specified each time). The fact that the research that I present mostly comes from Europe and North America may be interesting to many of the readers of this book because they may also come from those parts of the world. But at the same time, these people represent a minority of the people currently inhabiting our planet, about 12 percent of the total population according to an estimation,[9] and so one had better refrain from generalizing from this research. There are other specific problems relating, for instance, to extracting and analyzing DNA from fossils. Furthermore, researchers make any interpretations within particular theoretical frameworks that also need to be taken into account when one is considering their conclusions. All of these together produce evidence that supports, or not, particular hypotheses. Evidence becomes stronger only when there exist many studies in a field and researchers conduct robust meta-analyses in order to acquire a view of where the field as a whole is going.[10] Therefore, evidence is not something independent and absolutely objective, but depends on one's interpretation; it is, generally speaking, anything that can make a difference to what one is *justified* in believing.[11]

This being said, I should note that my main aim in this book is exactly to argue what the available empirical evidence makes us justified in believing about life. As I have already mentioned, robust developmental, historical, and evolutionary processes produce life outcomes. However, the details of these outcomes depend on particular critical events and are in no way predetermined. As I show in parts 2, 3, and 4, the outcomes of human development, human life, and human evolution are what they are because of particular critical events with particular outcomes in the context of broader natural processes; they are not the predetermined outcomes of fate, destiny, or design. This entails that the related beliefs that I present in part 1 are largely unjustified. Therefore, my aim in this book is to make readers appreciate the impact of critical events in life, against any notion of design, goal, or purpose.

In particular, I draw on developmental biology, history of biology, and evolutionary biology to show that human development, human life, and human evolution have a common underlying principle: *outcomes are not predetermined but are shaped by critical events.* Human development, life, and evolution are historical processes: they are sequences of successive events that are unique in space and time; that is, they took place at a certain time in a certain place—not anywhere, anytime (the philosophical term for this is that they are "spatiotemporally" unique). Thus, critical events can influence the course of the respective processes and make a difference in which one of several possible outcomes will materialize; which one this will be is previously unpredictable, but once it occurs there is a causal dependence of the future on that. We tend to think in terms of design and intentions because we usually consider only the actual outcomes but not the unrealized or currently nonexistent ones. This entails that we try to explain the outcomes in hindsight based on what we know that happened, and therefore the actual outcomes seem to us to be natural, predetermined, and even inevitable. However, the outcomes of human development, life, and evolution are neither predetermined nor entirely random; they are historically contingent.

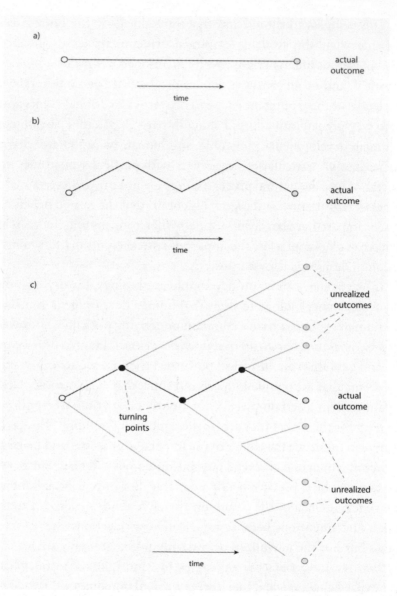

a)

actual
outcome

time

b)

actual
outcome

time

c)

unrealized
outcomes

actual
outcome

turning
points

unrealized
outcomes

time

Figure I.1. The sequences of events that we usually tend to perceive as being straight lines (a) are in fact not (b). In contrast, they are part of a larger ensemble of possible but unrealized sequences of events (c).

Let us explore the problem of hindsight a bit more. A main problem is that we tend to perceive outcomes as the terminal points of straight, linear sequences of events (figure I.1[a]). But, more often than not, things are not like this. We often fail to realize that the sequence of events that we are aware of may not proceed in a straight line but instead follow an indirect path (figure I.1[b]). Most important, we fail to realize that in parallel with the actual sequence of events, there were other possible sequences that never occurred. If that had been the case, they could have led to other outcomes. In other words, we fail to see that at certain points there were more than one possible outcome, and that critical events led to one or the other direction thus influencing which of the possible paths was actually taken. These points are the **turning points** that are the focus of the present book (figure I.1[c]). The sequence of events, of which historical processes consist, should be perceived as depicted in figure I.1(c) and not as in I.1(a).

Thirty years ago, paleontologist and prolific author Stephen Jay Gould wrote a book-length account of the significance of these turning points and of the impact of critical events on historical outcomes. That book was *Wonderful Life*, and the central concept therein was contingency, defined as the "affirmation of control by immediate events over destiny."[12] To illustrate the idea of contingency, Gould used the metaphor of the cassette tape: "You press the rewind button and, making sure you thoroughly erase everything that actually happened, go back to any time and place in the past . . . Then let the tape run again and see if the repetition looks at all like the original." If each replay resembled life's actual sequence of events, then one could conclude that whatever occurred somehow had to occur.[13] But this would not be the case, according to Gould, as "any replay of the tape would lead evolution down a pathway radically different from the road actually taken."[14] In his book, Gould focused on the findings of the Burgess Shale formation, a fossil-rich deposit in Canada with an exceptional preservation of soft tissues of animals. In this set of fossils one could see the details of organisms soon after the so-called Cambrian explosion some 570 million years ago, which was characterized by an enormous variation in forms and paved the way for the evolution of all major groups of multicellular animals. Gould concluded that the Burgess Shale fossils were an exemplar illustration of the impact of contingency in evolution. He thought that humans were no exception, and that we also are a contingent outcome of the history of life on Earth: "Replay the tape a million times from a Burgess beginning, and I doubt that anything like *Homo sapiens* would ever evolve again," Gould wrote.[15]

One of Gould's heroes in *Wonderful Life* was Simon Conway Morris. He was one of those who studied the Burgess Shale fossils and produced a new interpretation of the findings. However, he was not at all in agreement with Gould's conclusions about the impact of contingency on the evolution of life on Earth. In contrast, he later argued that what was important was the likelihood of the emergence of particular properties. Certain properties have appeared again and again in the evolution of life: "The tape of life, to use Gould's metaphor, can be run as many times as we like and in principle intelligence will surely emerge."[16] Conway Morris argued that several similar characteristics have independently evolved at different times in different lineages, a phenomenon called *convergence*. Similar characteristics can evolve independently in different lineages, if they are advantageous. One such example are the wings of birds and of bats, which serve the same function but are structurally different. In particular, the wings of birds consist of their fore-limbs whereas the wings of bats consist of their elongated digits that are connected via a webbed membrane of skin. According to Conway Morris, evolution is only possible in particular directions, not in any direction; in other words, the number of evolutionary pathways available to life are rather limited, not endless. At the same time, similar environmental pressures can favor particular advantageous characteristics and not others. In this sense, humans would have evolved on earth in some way or another.[17]

In the present book I consider the impact of contingency not only in human evolution but also in human life and development. Based on these considerations, I argue (i) that the same principle, that critical events shape outcomes, underlies human development, human life, and human evolution, and (ii) that the same human intuitions preclude us from realizing this. Thus, whereas accounts for particular historical outcomes should cite particular antecedent conditions and highlight the impact of particular contingencies, we often tend to prefer accounts that present these outcomes as fulfilling some final end. A major characteristic of explanations of historical outcomes is that they are narrative explanations. Gould argued that the explanations of certain evolutionary outcomes (and I would add developmental and life outcomes as well) can take the form of a historical narrative. This would explicitly mention the contingencies of the antecedent states, which, had they been constituted in a different way, these outcomes would have not been brought about. These contingencies do not diminish the explanatory power of the narrative; a historical explanation can reach the same level of confidence as any physical explanation under invariant laws of nature, such as those in physics, insofar as enough details about the antecedent states are available in order to understand their causal relation to the observed outcome.[18] This means that we can adequately explain outcomes of historical processes such as human development, life, and evolution insofar as we have enough information about the antecedent conditions and the critical events that brought about these outcomes. Of course, this information is not always available.

Philosopher John Beatty has carefully and diligently analyzed the evolutionary contingency thesis and has distinguished between two versions, which Gould himself did not clearly distinguish: unpredictability and causal dependence. In my account in the present book, I consider these as two complementary aspects of contingency and not as two different versions. Indeed, Beatty himself noted that "we might even think of them as complementary components of a combined interpretation, according to which: a historically contingent sequence of events is one in which the prior states are necessary or strongly necessary (causal-dependence version), but insufficient (unpredictability version) to bring about the outcome."[19] Let us consider these two aspects of contingency in more detail, by means of an illustration. Imagine that once an event A occurs, there are two possible outcomes: B and C. Whether it will be B or C that will occur after A is previously unpredictable; B and C could be equally probable, or one of them might be more probable than the other. In either case, insofar as one cannot tell in advance which one of them will actually occur after A, both B and C are contingent *per se*. Assuming that event B occurred, it was neither necessary nor bound to occur (this is the unpredictability aspect of contingency). Now, subsequent events B1 and B2 are contingent *upon* B, meaning that their occurrence depended on the occurrence of B. Assuming that it was B2 that actually occurred, its occurrence causally depended on whether or not B occurred—because it would not have occurred if B had not previously occurred (this is the causal dependence aspect of contingency) (see figure I.2). In the same sense, event C was contingent *per se*, but did not actually occur; neither did C1 nor C2 occur, because they were both contingent *upon* C. Therefore, events like B are critical because they may determine which of several possible paths will be followed. Such events that are *contingent per se*, and that subsequent events are *contingent upon* them, are called **turning points**.[20] In the present book I argue that narrative explanations are appropriate for explaining the outcomes of human development, human life, and human evolution, and that turning points are central in these explanations. I also explain why we should not perceive sequences of events such as A-B-B2 as predetermined or inevitable, and that we ought to also consider the possible but unrealized outcomes such as B1, C, C1, C2 (figure I.2).

Let us consider a more concrete example of a turning point, by analogy with a simple physical process. Imagine that you release a light-gray ball on a plane from which there are three possible routes: A, B, and C, each of which leads to the respective endpoint a, b, and c. These three routes are equally probable to follow, because the plane is designed and constructed in such a way that does not bias the direction a ball will take. Imagine now

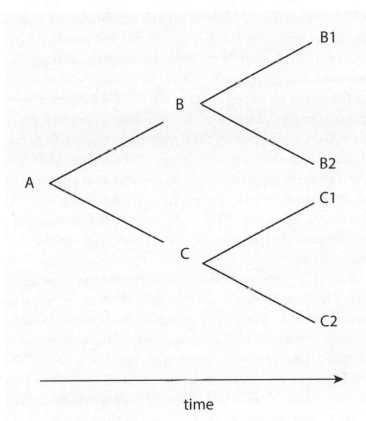

Figure I.2. An illustration of turning points. Assuming that the actual sequence of events was A-B-B2, B is a turning point because it was *contingent per se* (C and not B might have occurred—unpredictability aspect of contingency) and subsequent events were *contingent upon* it (B2 occurred only because B occurred earlier—causal dependence aspect of contingency). The problem is that, intuitively, we tend to see only the historical sequence A-B-B2 and to overlook the possible but unrealized outcomes (B1, C1, C2).

that at each of these endpoints there is a bucket full of paint: white in a, dark gray in b, and black in c. Finally, imagine that the ball will be completely painted as soon as it lands inside each of these buckets. Now here is the impact of contingency: whereas it is not possible to tell in advance which of the three routes A, B, or C the ball will follow (unpredictability aspect), the color that the ball will have in the end (white, dark gray, or black) will entirely depend on the route taken, as a bucket with paint of a different color is found at the end of each route (causal dependence aspect). No matter how many times we do this experiment, it will always be impossible to predict which of the three routes the ball will take, and its color will always depend on the route taken (figure I.3). What is the critical factor? It is which one of the routes A, B, or C the ball will follow, which in turn depends on how it will be released on the plane and how it will roll on the plane until it reaches one of the holes that leads to one of the three routes. Taking any of the three routes A, B, C is *contingent per se*; and the color that the ball will eventually come to have will be *contingent upon* the route taken.

Processes like these are often used to illustrate the concept of randomness, which is relevant but which should be clearly distinguished from contingency. In statistics, a sequence of numbers is random if it is impossible to predict the successive values. Therefore, randomness is about unpredictability in a sequence of events. In the case of the balls in figure I.3, we can say that which route the ball will follow is entirely random. What this means is that if you release three balls consecutively on the plane, you cannot tell in advance what the sequence of routes taken will be. There are twenty-seven such possible combinations (we have three objects combined in groups of three, so the possible sequences are 3^3, or $3 \times 3 \times 3 = 27$), which are presented in table I.1. The three balls could take each a different route; two balls could take the same route; or all three balls could take the same route. Now, before you have a look at table I.1, let's address the question: If you performed the experiment of releasing the three balls six times, which of the twenty-seven possible outcomes (e.g., A,B,C; B,C,A; etc.) would be more probable? You can write down your guessed path for each of the six drops: (1) _____ (2) _____ (3) _____ (4) _____ (5) _____ (6) _____.

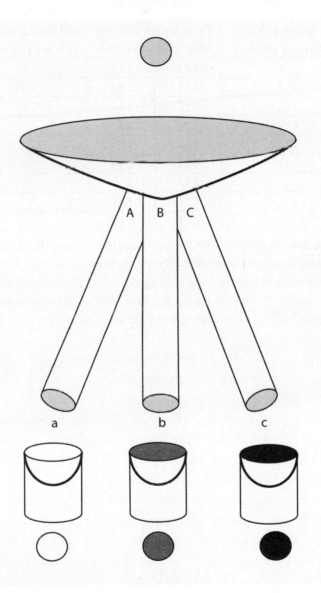

Figure I.3. If we release the ball on the plane, we cannot predict which of the three routes (A, B, C) it will take (unpredictability aspect of contingency); but the color it will come to have (white, dark gray, black) will depend on the route actually taken (causal dependence aspect of contingency).

All balls taking different routes	Two balls taking the same route		All balls taking the same route
A B C	A A B	C C A	A A A
A C B	A A C	C B B	B B B
B A C	A B B	C A A	C C C
B C A	A C C	A B A	
C B A	B B A	A C A	
C A B	B B C	B A B	
	B A A	B C B	
	B C C	C A C	
	C C B	C B C	

Table I.1. The twenty-seven possible combinations of routes taken when three balls are released consecutively on the plane of figure I.3. Each letter corresponds to the route taken by a ball. In all cases, the first letter corresponds to the route taken by the ball that was released first; the second letter to the route taken by the ball that was released second; and the third letter to the route taken by the ball that was released third.

So, what do you think? If you have mostly written the combinations in the left column, having thought that it is more probable for the three balls to take different routes because each route has the same probability . . . I am sorry, you are wrong![21] It is actually more probable that two of the balls will follow the same route. To put it simply, there are six different ways that each of the three balls can take different routes, but eighteen possible ways of two balls taking the same route.[22]

Now when it comes to historical processes like development, life, and evolution, the outcomes are not totally random but are biased in a sense, because they depend on what has happened before.[23] For example, if the apparatus was redesigned such that route C was possible only after going through route B, then the probabilities for taking each of these routes would not be independent—because the two events are not independent. Only those balls that would take route B would also take route C (see figure I.4). This also nicely illustrates the idea of causal dependence: a ball can take route C only if it also takes route B; but a ball that takes route A will never take route C. What I want to note here is that the outcomes of historical processes are not entirely independent, and we cannot simply calculate their probability through the probabilities of the various events. An event may generally be of low probability, but it is almost certain once another event takes place. This means that in historical processes, which are sequences of events in which one brings about the other, we need to consider the probabilities of all the evens in a sequence in order to estimate or explain the probability of the final outcome.

To sum up: critical events shape outcomes by influencing the direction of a process toward a particular path among several possible paths. Which of these will be followed is previously unpredictable, but once taken the outcome depends on it. In *Turning Points*, I argue that this causal dependence often makes us in hindsight perceive outcomes in our development, lives, and evolution as inevitable. This we do because in hindsight we selectively pick up past events and use them to explain these outcomes as inevitable, overlooking the impact of critical events that were turning points. Yet, I argue, many of these outcomes were evitable, because they were causally dependent upon unpredictable critical events. Our development, life, and evolution could have thus taken other paths, resulting in different, alternative outcomes than those that actually occurred.

These alternative outcomes are often described as counterfactuals. Counterfactuals can be defined as "alternative versions of the past in which one alteration in the timeline leads to a different outcome from the one we know actually occurred."[24] Imagining these alternatives versions of the past that could bring about different outcomes is crucial for realizing the importance of turning points. One does not need to figure out exactly what these different outcomes could be, but only consider their plausibility. One might think that imagining and describing alternative worlds are appropriate for novels and for books on science fiction, as we cannot really know—but only imagine—how these alternative worlds could have been. This is indeed true for many cases. However, there are several cases in which this is possible. For instance, consider the personal

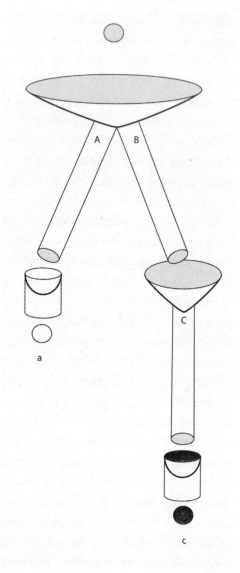

Figure I.4. If we release the ball on the plane, we cannot predict which of the two routes (A or B) it will take (unpredictability aspect of contingency); but the color it will come to have (white or black) will depend on the route taken, and it can only take route C after having taken route B (causal dependence aspect of contingency).

story that I described in the preface. Had my father died in July 1974, my brother would not have been born, and this is a counterfactual outcome that can be identified with certainty. I am also certain that my whole life would have been a lot worse than what it was, had I grew up without my father around, whatever my mother and the rest of the family could have done for me. But this is a counterfactual outcome that I can identify with less certainty, because I cannot really know. For instance, my mother could have raised me with love and affection that would have compensated the love and affection that my father actually provided me with. I seriously doubt this, as the loss of my father would probably have stigmatized my life forever, but I accept that in this case I cannot really know. However, the important issue here is that, in hindsight, I do not take for granted that my father lived. I am especially sensitive to the fact that his survival was contingent *per se*, and the lives of my brother, our mother, and myself were contingent *upon* that.

My main aim in *Turning Points* is to highlight the contingent character of actual sequences of events that some people take for granted in hindsight. It is the certainty of hindsight that actually blinds us in seeing the possibility of counterfactual outcomes. The objective, therefore, is to liberate ourselves from hindsight and try to see past events and historical processes as open to various, unpredictable possible futures. Whatever these outcomes could be, considering their possibility and plausibility can help us grasp the importance of turning points in the context of robust developmental, historical, and evolutionary processes. To achieve this, we need to adopt a new perspective that is free from hindsight and that considers events in the context in which they actually occurred, and not after they did. It is only then that we can begin to consider that there could be alternative possible paths (see figure I.5). Therefore, my goal is to help you replace the perspective in figure I.5a with that in figure I.5b—or to facilitate you to help others do this. In doing so, you might come to see more clearly that the actual outcome has been what it is because of the path taken at particular turning points, whatever the possible alternative outcomes could have been. You do not need to know the details of these alternative outcomes; you only need to recognize that alternative outcomes were possible.

A relevant distinction is that between multifinality and equifinality. Multifinality is the idea that an outcome is a part of a complex causal process, in which different antecedent conditions could produce different outcomes. Therefore, an outcome that actually took place might have not taken place under different antecedent conditions. For instance, as I also described in the preface, I met my wife because she was invited by a person I would be meeting on that day to join us, and she accepted. If she had not been invited, or if she had not accepted, we might have never met. As a result, we would not have gotten married and our family would not have existed. In contrast, equifinality is the idea that an outcome is somehow fixed and meant to happen. Therefore, different antecedent conditions are equally likely to produce the same outcome. For instance, according to this view, if my wife and I had not met on the particular day that we actually met, we would have probably met at some other time and eventually we would get married and start our family. When we look at outcomes with hindsight, equifinality may seem intuitive because we may be able to think about several antecedent conditions that might lead to the same outcome—in my example, having that person invite my wife to join us at some other time, or re-invite her even though she declined the first time, and have her join us then. When we take a certain outcome as given, it is possible for us to think of various alternative conditions that might have led to that—for instance, thinking that because my wife and I had a common acquaintance, we would have met somehow. The problem is that in hindsight we take the actual outcomes as given, and we think that they would somehow have occurred anyway, and perhaps even consider them inevitable. But if we manage to liberate ourselves from hindsight and look at the events as they took place, we may realize that multifinality is more likely and that things might have easily turned out differently than how they actually did.

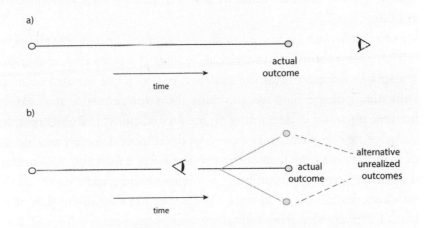

Figure I.5. (a) When we look at events with hindsight, after an event has occurred, we only look backward in time and only see the path actually taken and the actual outcome. Thus, we may think of this outcome as inevitable. (b) We need to imagine ourselves within the actual context of past events, before the actual outcome that we observe took place, in order to imagine the possible alternative, unrealized outcomes.

In part 1, I present prevalent conceptions about fate, destiny, and intelligent design, which act as obstacles to laypeople's understanding of the living world. These obstacles are then challenged in the subsequent parts 2, 3, and 4. Researchers in psychology, science education, and the public understanding of science have concluded that several unscientific conceptions are quite widespread among people. These notions share the common underlying idea that external or internal factors, which humans are entirely unable to control, determine the outcomes in human development, life, and evolution. In this sense, these outcomes may be considered predetermined and inevitable. Interestingly, this idea feels very intuitive to humans. In part 1, I offer evidence that these unscientific conceptions are quite prevalent, and I explain how and why people intuitively respond or react to tasks used in research. In many cases, I also provide you, the reader, with the opportunity to take these tasks before reading further, in order for you to reflect upon what you think—or what you think that you think.

One important aspect of human development is that it is characterized by both robustness (individuals exhibit the general characteristics of a species irrespective of the environment they live in) and plasticity (individuals of the same species with the same genotype may exhibit different phenotypes depending on local conditions).[25] Robustness and plasticity are complementary aspects of development, yet we tend to pay more attention to the former than to the latter. However, the outcome of development depends on both our genome and critical events during our development. Based on the cases discussed in the second part of the book, I explain why genetic fatalism cannot account for human development and for the origins of traits and disease. The characteristics of humans are the outcome of critical events that took place during their development. Whether one will be born at all, as well as which traits one will exhibit depends on various critical events and is not predetermined in one's genes. Besides highlighting the criticality of reproductive and developmental events, another aim I have in part 2 is to teach some basic biology. Most people should remember learning about mitosis and meiosis at school, however I guess that many of them must have wondered at that time why they needed to learn this stuff. As I show, many events that occur during meiosis, the division leading to the production of spermatozoa (sperm) and ova (egg), have a big impact on how humans develop—if they develop at all. Mitosis, the division by which cells proliferate and an embryo grows, can also have a big impact on its development.

In the third part I aim at showing that there is no inevitable destiny in life but people do whatever they do because of critical events and their decisions and actions related to them. As an example, I show that important advancements in our understanding of life were not waiting to be made, and those who made them were not destined in any way to do so. The contributions of particular individuals made a difference, and these in turn were contingent on the conditions under which these individuals lived and worked, and on particular critical events. Charles Darwin's life and theory serve as a case study here. Darwin's theory had the form and content it had in *On the Origin of Species* and it was published in 1859 because of particular turning points: the *Beagle* voyage between 1831 and 1836; Darwin's reading of Thomas Malthus in 1838; the publications and the public reception of the *Vestiges of the Natural History of Creation* in 1844; and Darwin's reception of the letter by Wallace outlining his own theory in 1858. Darwin's theory could have been published in a different form, had it been published earlier or later than 1859. However, particular turning points led to the publication of *On the Origin of Species*, which was far from inevitable. I must note that these are not the only turning points in Darwin's life but the ones that my reading, and interpretation, of his autobiography made me conclude that he considered as the most important ones. I must note, also, at this point that I used the development of Darwin's theory and not a broad historical event—such as how World War I started, to take a classic example—because it is at the individual level that the impact of contingencies becomes clearer.

Finally, in part 4 I show the impact of contingent events on human evolution. Based on the cases discussed in that part, I explain why intelligent design cannot account for the emergence of humans. Human features were not designed but are the outcome of evolutionary processes. The distinctive characteristics of humans discussed in those chapters are the outcome of critical events that took place during human evolution. How we have evolved to be was far from inevitable and the outcome of contingent events. This entails that neither our presence in this world was inevitable. Contrary to the view that the conditions in our world have been appropriate so that life could arise (and/or evolve), I argue that life has evolved because the conditions have been those that allowed this to happen. Had the conditions been different, it is possible that life as we know it and humans might have never existed.

The present book is intended for the general reader who is interested in biology and in understanding how turning points have made us who we are. It brings together research from psychology, science education, developmental biology, history of biology, and evolutionary biology to show that—contrary to popular, intuitive views such as genetic fatalism, destiny, and intelligent design—it is critical events that shape outcomes. My aim is to show that whereas people tend to intuitively perceive purpose and design in nature, contingency plays an important role in human development, life, and evolution. For this purpose, scientific evidence is used to show why the idea of purpose and design in human development, life, and evolution is implausible. This is presented after a review of the evidence that several people do indeed tend to have such beliefs. Finally, I also aim at making the general reader familiar with some fundamental knowledge and contemporary research on the respective fields.

I should note that I do not assume that all readers think that there is purpose and design in nature, although this is the case for many people. However, I have seen that even well-informed people find hard to accept that there is no design and no purpose in nature because such a view conflicts with their intuitions. Therefore, a main target of the present book is the intuitiveness of this idea. Those who see plan and purpose in nature but are open to reconsidering their own views might find this book useful for reflection. And those who do not see plan purpose in nature might find this book useful for getting arguments and evidence to help those who think otherwise to understand why this might not be the case. I think that this is important because I have seen several people who believe in fate, destiny, and design both be passive and indecisive and accept events as they come, without ever trying to change anything. Of course, we cannot change our development and evolution; but if we appreciate the impact of contingencies, we might try to change our lives by at least refraining from holding unjustified beliefs.

I have focused, here, on humans in order to make this point easier to explain. Writing about human development, life, and evolution inevitably draws on examples that are "close to home," about which people have most likely wondered and which they might find easier to understand. In addition, I think that this most likely makes the book more interesting; because it is about us! We, humans, are nothing more than a very recent and very short branch in the enormous web of life on Earth—an explicit disclosure from my part, if you think that I am being unnecessarily anthropocentric. Any such book about any such organism could be important, be interesting, and provide valid arguments for the role of critical events in development, life, and evolution. But understanding better the human condition is something that many of us have thought about—how many of you have sincerely wondered about the evolution, development, and lives of species such as *Daphnia* or *Drosophila*? My guess is very few of you. Therefore, I hope that my focus on humans, albeit a narrow one, has indeed made this book both intelligible and interesting to you. Because it is about you.

A final point. Many events can be unpredictable and have several possible outcomes, only one of which will materialize. In this sense, many events could qualify as turning points. This raises the question: are all these "real" turning points? If yes, then are there turning points that are more important than others? In my view, the criteria for identifying a "real" turning point stem from the definition of a turning point itself. I defined a turning point as an event that is unpredictable (contingent *per se*, because other events were also possible but it was that and not the others that occurred) and that has significant consequences (future outcomes are contingent *upon* it, because that event was necessary for the subsequent evens to occur). Many events are unpredictable but do not have significant consequences; also, many events have significant consequences but are relatively predictable. A turning point is thus an event both that is unpredictable (in terms of whether and how it will occur) and that impacts what comes next (in the sense that the future will depend on how that event occurred). Whereas the criterion of unpredictability is more or less clear—something that one could not have foreseen—one might wonder about the criterion of significant impact. An unpredictable event becomes a turning point when its impact is significant, either in terms of time (a long-term impact), or in terms of quality (because it narrowed down the potential features and thus shaped the outcome). The examples that I give in parts 2, 3, and 4 of this book have both of these features. I should note again that these are not the only turning points in human development, Darwin's life and publication of his theory, and human evolution. But I argue that all of these are genuine turning points that provide evidence that thinking in terms of fate, destiny, and intelligent design in unjustified.

Let us now explore this thinking in more detail.

Part 1

THE DESIGN STANCE

Chapter 1

"WHY X?": "IN ORDER TO Y"

Have you ever asked yourself why we have hearts? Most people would reply: "In order to pump blood." Many human characteristics seem to serve goals, and these goals look quite obvious in many cases. Why do we have legs? "In order to stand up and walk" is an answer that would make sense to most people. Why do we have opposable thumbs? "In order to grasp and handle objects," most people would also reply. And this can go on for any of our body parts and organs you can think of: brain, stomach, lungs, liver, to name a few. The roles of other characteristics may seem less obvious, but they also seem to be there. Why do we have eyebrows? "In order to prevent sweat from entering our eyes" is a plausible answer. Why do we have hair on our heads? "In order to keep our brain warm" is again a plausible answer. But if you think harder, it is not easy to assign goals to all characteristics. Why do we have toes? Why do men have nipples and facial hair? And so on. Perhaps there is a less obvious role in these cases, *but it must be there*, you may think. Otherwise, if they do not do anything, why would we have certain characteristics? In other words, why would characteristics that serve no use or purpose exist at all? You may have thought about questions like these in the past, and you may have arrived at certain answers like those above. In this chapter I invite you to forget for a moment your earlier conclusions and follow me on a foray into thinking about purpose and design in nature. Answers to such "Why X?" questions that explain the presence of a characteristic are considered explanations. Throughout the present book, I consider explanations as statements that identify the causes of a biological characteristic or phenomenon, that is, that provide an account for why it happened or why it came to be. And when these explanations take an "in order to Y" form, they are described as teleological explanations (*teleological* means that their logic is based on the goal—*telos*—that they serve).[1]

The discussion about purpose and design in Western culture goes back more than two thousand years, at least back to Plato and Aristotle. Plato believed that the universe was created by a Divine Craftsman, the Demiurge (Creator). Plato considered the transfusion of the soul of the Demiurge into the universe as the final cause of its creation. This process had to take into account the actions of Need, the mythical equivalent of the properties of the structure of matter, which imposed constraints to the work of the Demiurge. Plato thus recognized two types of causes, which he viewed as interdependent and not in conflict. Therefore, the universe was an artifact that resulted from the purposeful and rational action of the Demiurge that had dominated over the irrational Need. This is a view of the world being "unnatural," in the sense that it is the product not of natural processes but that of a divine craftsman. Aristotle was a student of Plato who tried to answer questions about phenomena in organisms by looking for natural causes. He thus described four causes that acted in nature: the *efficient* cause, the *material* cause or *matter*, the *formal* cause or *form*, and the *final* cause; and he considered all four of them necessary for understanding. The classic illustration of these causes is with the example of a statue: (1) the *material* cause is the bronze of which the statue is made, which undergoes a change that results in the statue; (2) the *formal* cause is the shape of the statue, as the bronze is melted and used in order to acquire a particular shape; (3) the *efficient cause* is the knowledge that the craftsman implements to create the statue; and (4) the *final cause* is the goal for the fulfillment of which the whole process of the production of the statue is taking place. Even though the example of the statue might make one think that Aristotle thought in terms of design like Plato, this is not the case. For Aristotle, there were no external intentions, and the final causes served the maintenance of an organism. Thus, the final cause for the existence of an organ would be its usefulness to the organism that possessed it and not, like Plato, the intentions of a divine designer.[2]

This outline of the Platonic and Aristotelian teleology encapsulates the essence of teleological explanations. On one hand, there is the view that teleology is external to the entity under discussion; the final ends are determined by a conscious, external agent who intentionally designs something with a purpose in mind. Details notwithstanding, this kind of external teleology can be illustrated with the barbs of barbed wires. Humans designed and created barbed wires in order to protect something that is of value to them. In this sense, the barbs of the barbed wire exist in order to fulfill a purpose that is external to the barbed wire itself. On the other hand, there is the view that teleology is internal to the entity under discussion; the final ends are determined by the usefulness of particular features for the entity that bears them itself, and not by any conscious, external, intentional agent. Details notwithstanding, this kind of internal teleology can be illustrated with the thorns of roses. These protect the roses themselves, for instance, from animals that might want to eat them. In this sense, the thorns of roses exist in order to fulfill a purpose that is internal to them, their survival, and not determined by any external agent. Therefore, we can state that both barbs and thorns exist for a purpose, but this purpose is different: it is external in the case of barbs and internal in the case of thorns. These are two main types of teleological explanations. In this chapter, as well as throughout the whole of part 1, I explore the intuitiveness of teleological explanations in various contexts by presenting the findings of studies about how humans rely on such kinds of explanations in order to account for the development and evolution of biological traits, as well as of life outcomes. In this chapter, I first present research about the prevalence of teleological explanations in general. Then, in the subsequent chapters, I present explanations related to human development, human life, and human evolution. I should note that I do not always get into the details of whether the teleological explanations under discussion are of the "external" or the "internal" type.

However, whenever it is really important to be aware of this for better understanding how people think, I mention this.

Several studies have shown that explaining the presence of particular characteristics in terms of design and purpose characterizes our thinking from very early in childhood. In one study, children were asked to choose between two possible explanations for why an object had a certain property. You can try it yourself:

1. Two people are talking about why plants are green.
    (a) One says it is because it is better for the plants to be green and it helps there be more plants.
    (b) The other says it is because there are tiny parts in plants that when mixed together give them a green color.
    Which reason is a better one?

2. Two people are talking about why emeralds are green.
    (a) One says it is because it is better for the emeralds to be green and it helps there be more emeralds.
    (b) The other says it is because there are tiny parts in emeralds that when mixed together give them a green color.
    Which reason is a better one?

You can write down your choice here for 1. _____ and 2. _____. Did you choose (a) or (b) in each case?

In both (1) and (2), answer (a) is a teleological explanation, as green color is considered to serve a purpose; in contrast, (b) is a physical explanation that explains the property of being green on the basis of what the object consists of. Did you think that the teleological explanation works better for plants, whereas the physical one works better for emeralds? This is how second-grade and kindergarten children from the United States, who participated in that study, responded. They preferred teleological explanations for organisms, such as plants, over physical explanations.[3] In contrast, the latter were preferred for nonliving objects, such as emeralds, a lot more than teleological explanations. Studies like this one have shown that children prefer teleological explanations for organisms compared to nonliving natural objects. How about organisms and artifacts? There is evidence that although they think of both organisms and artifacts in teleological terms, children as young as three years old can distinguish between their self-serving and other-serving roles. For example, as already mentioned above, roses can be considered as having thorns to keep animals from getting at them, thus supporting their survival, whereas barbed wire can be considered to have barbs in order to prevent animals from accessing something that is valuable to humans.[4]

In another study, kindergarten-aged school children and university undergraduates in the United States were shown twenty-two magazine photos of various objects (see table 1.1).[5] As each photo was presented, the researcher asked: "What's this for?" while pointing to either the object or a part of the object (e.g., "What's the tiger for?" or "What's the tooth for?"). All participants were explicitly given the option of replying that either the object or the part was not "for" anything. You can take the test yourself; write down your answers on the table on the next page before reading below!

Let us see . . . How many "in order to" answers did you give? If you did this for more than half of the entities and their parts in table 1.1, then you are thinking in teleological terms. For instance, a teleological response to the question "What's the clock for?" would be "In order to tell the time." Results supported the conclusion that whereas there were no significant differences between children and adults in providing teleological explanations for the parts of organisms, for the parts of artifacts, and for whole artifacts, there was a significant difference in that children provided more teleological explanations for whole nonliving natural objects, parts of nonliving natural objects, and whole organisms, when compared to adults. Overall, adults viewed, for example, clocks, pockets, and beaks as being "for" something but did not think that lions, mountains, and the parts of mountains have functions. In contrast, children

Entity	What is it for?	Part	What is it for?
woman		hand	
man		ear	
baby		toes	
little girl		fingers	
lion		leg	
tiger		tooth	
plant		leaf	
tree		trunk	
cow		udder	
cat		paw	
baby bird		beak	
puppy		mouth	
clock		hand	
jeans		pocket	
ring		stone	
statue		arm	
mountain		protuberance	
cloud		trail	

Table 1.1. Some entities and their parts. Can you think of what each of them is for? Get a pencil! (Note that each part belongs to the entity on the same line.)

generally attributed functions to animals and their body parts, to clocks and their components, and to nonliving natural objects and their parts. For instance, some explanations were that birds exist "to fly," plants "to grow," mountains "to climb," and clouds "to rain." In general, children appeared to view all kinds of objects as existing "for" something but were more likely to ascribe a function to their parts, for example, to a tiger tooth rather than to the whole tiger. The inference that can be made from these findings is that at young ages we have a broad tendency to think in teleological terms for all kinds of entities, then restrict it to artifacts and organism parts in adulthood. In other words, it could be argued that a generalized teleology is the default human stance, and that it is later appropriately restricted as we come to learn more and better understand the world.

This was shown even more clearly in a different study that aimed at investigating what kinds of properties children and adults considered as "designed for a purpose," and what kinds of teleological explanations they would find acceptable.[6] University undergraduates and children aged seven to ten years from the United States participated in this study. Each time, they were shown realistic pictures of an unfamiliar prehistoric animal and an unfamiliar nonliving natural object, for example, an aquatic reptile and a pointy rock. Participants were asked to choose between a physical and a teleological explanation for: (a) a biological property of the animal (e.g., long neck); (b) a behavioral property of the animal (e.g., neck swayed from side to side); and (c) a property of the nonliving natural object (e.g., pointy). For example, for the question about why certain organisms had long necks, participants had to choose between the physical explanation ("They had long necks because the stuff inside got all stretched out and curved") and the teleological one ("They had long necks so that they could grab at fish and feed on them.") For the question about why there were pointed rocks in the area in which those organisms lived, participants had to choose between the physical explanation ("They were pointy because little bits of stuff piled up on top of one another over a long time") and the teleological one ("They were pointy so that animals wouldn't sit on them and smash them"). The researchers concluded that adults thought of biological and behavioral properties as existing to serve purposes for the animals but did not accept teleological explanations about nonliving natural objects such as pointy rocks. However, children of all ages not only accepted teleological explanations of the biological and behavioral properties of organisms but also preferred teleological over physical explanations for objects such as pointy rocks. Once again, the inference that can be made is that a generalized teleology seems to be prevalent in young children, whereas a more restricted one is found in adults.

In another study, American children were shown eight photographs of two artifacts (a hat and a boat), two animals (a bird and a monkey), two natural events (a thunderstorm and a flood), and two nonliving natural objects (a mountain and a river); they were then asked three different kinds of questions.[7] These questions took three different forms: (i) open-ended questions such as "Do you know what a thunderstorm is?" and "Why did the first ever thunderstorm exist?"; (ii) closed-ended questions for the origin of the object or phenomenon, in which they had to choose between two alternative explanations, a physical one ("The first ever thunderstorm occurred because some cold and warm air all rubbed together in the clouds") and a teleological one ("The first ever thunderstorm occurred to give the earth water so everything would grow"); and (iii) intelligent design questions of the form "Did someone or something make the first ever thunderstorm exist, or did it just happen?" From the results it was concluded that in the open-ended questions, children provided more teleological explanations for artifacts than for animals or nonliving natural objects. They also gave more teleological explanations for all of these categories than for natural events. For the natural events, in turn, more physical explanations were given; such explanations were never given for artifacts and only rarely for animals and natural objects. In the case of the closed-ended origins questions, children selected teleological explanations more for artifacts than for animals or nonliving natural objects, and even less for natural events. Finally, in the case of the intelligent design questions, more children thought that someone had caused artifacts to exist than they thought someone had caused animals to exist; and they thought even less that someone had caused nonliving natural objects or natural events.

Studies like these overall support the conclusion that children tend to think in teleological terms about organisms, artifacts, and nonliving natural objects, and that teleological explanations are restricted to organisms and artifacts in adulthood. Leaving nonliving natural objects aside, children and adults seem to think of organisms, artifacts, and their parts in teleological terms. Does this mean that they think of organisms as artifacts designed for a purpose? Not necessarily. For instance, in another related study, American children were shown pictures of eight unfamiliar animals and eight novel artifacts, and they were told that they could ask any questions about them.[8] Children asked an average of twenty-six questions each, most of which concerned whole objects, whereas some concerned their parts as well. Some questions were appropriate for both organisms and artifacts, for example: "What is it?" or "Where do you find it?" Questions about intended function and use, such as "What is this for?" or "Is this for doing X?" were asked a few times for artifacts but not for animals. Questions about eating habits and reproduction were also asked for animals, but not for artifacts. Finally, it was found that children asked more function-related questions about animal parts than for whole animals. In contrast, there was no difference in the proportion of function-related questions asked for artifact parts and whole artifacts. However, the proportion of function-related questions about artifact parts and organism parts were quite similar (a point to which I return below). The researchers concluded that young children look for different information about animals and artifacts without thinking about organisms in terms of design and function. This conclusion stands in contrast to the research presented above that children provide teleological explanations for all kinds of entities, and think of both organisms and artifacts in terms of design.

Despite this difference, among all of the studies presented so far, there are two common findings that should be noted. First, children generally provided teleological explanations for, or asked questions about, the parts of organisms and artifacts. Even if children perceive animals as being different from artifacts, this does not mean that they necessarily perceive animal parts differently from artifact parts. It actually seems plausible that our understanding of artifacts influences our understanding of organisms. From very early in our lives, most of the objects that we encounter are artifacts, and we immediately see that they are used in order to achieve some goal. Just think of how many artifacts were around you when you were an infant; you may have spent most of your first few years of life inside your home, surrounded by toys, feeding bottles, puppets, spoons, plates, chairs, tables, and numerous other artifacts. How many animals did you encounter as an infant? A lot less I assume. Did you pay any attention to nonliving natural objects such as clouds or rocks? Perhaps, but artifacts around you were many more. This is not the case only for children growing up in cities. Even if you grew up in the countryside and your parents were farmers, the animals and the nonliving natural objects that you encountered were probably a lot less in number and variety compared to the artifacts around you. Therefore, it could be the case that our understanding about the intentional production and use of artifacts from a very young age makes us extrapolate this way of thinking to nature. This does not necessarily mean that we perceive organisms as designed for a purpose; but it could be the case that we perceive the parts of organisms as being there for a purpose—a function—as it is the case for the artifact parts. For example, because we see that the wings of airplanes are for flying, we might reasonably presume that the wings of birds are for flying too. We thus tend to explain the existence of parts on the basis of what they do. At the same time, thinking about objects in terms of purposes and functions also has a heuristic value for better understanding and using them. This means that asking what something exists for has a practical value in guiding us to wonder what kind of positive contribution this thing might have and therefore perhaps discover its function. This may enhance our tendency to ask teleological questions for the parts of artifacts and organisms (like the back of a chair, the teeth of a tiger), even though we may not ask such questions for nonliving natural objects (like pointed rocks).

Although the studies discussed so far involved people from the United States, there is evidence that supports the conclusion that teleological biases are perhaps universal. A set of two studies with children and adults in China aimed at investigating whether people from a non-Western country showed the same preference for teleological explanations as people from Western countries have been found to do. In particular, researchers investigated whether Chinese children exhibited a preference for teleological explanations for the properties and origins of natural phenomena, and they also compared their explanations to those of Chinese adults. In the first case, similarly to the study with American children discussed above, four ancient animals and objects in their habitats were shown to participants and they were asked why they had particular properties (e.g., long necks, pointy rocks), asking them to choose between a teleological explanation (e.g., "The rocks were pointy so that animals wouldn't sit on them and smash them") and a physical one (e.g., "The rocks were pointy because bits of stuff piled up on top of one another for a long time"). In the second case, participants were asked an open-ended question about the origins of animals, natural objects, natural events, and artifacts; and then they were asked to choose between a physical explanation (e.g., "The first ever thunderstorm occurred because some cold and warm air all rubbed together in the clouds"), and a teleological one (e.g., "The first ever thunderstorm occurred to give the earth water so everything would grow"). The results showed that overall children endorsed teleological explanations for different kinds of natural phenomena, which decreased across the grades, whereas adults mostly endorsed teleological explanations for cases that were scientifically legitimate, such as for organisms and artifacts. Therefore, a conclusion that can be drawn is that the preference for teleological explanations may not be a product of Western culture but rather a basic aspect of human cognition.[9]

Even though the results of the studies presented so far support the conclusion that adults and children differ in how they attribute functions to objects and their parts, researchers wanted to examine in more detail adults' teleological explanations. In one study, 109 American undergraduate students were divided in three groups, were shown explanations for "why things happen," and were asked to decide whether these were correct or incorrect. Students in the first group were shown each explanation on a screen for 3.2 seconds (fast-speeded group); students in the second group were shown each explanation on a screen for 5 seconds (moderately-speeded group); and students in the last group read the explanations directly on the answer sheets without any time limits (un-speeded group). Table 1.2 presents some of the explanations shown to participants, and you can actually test yourself; it will not take more than five minutes.

Explanations	Correct (C)/Incorrect (I)
1. Earthworms tunnel underground to aerate the soil	
2. Mites live on skin to consume dead skin cells	
3. Mosses form around rocks to stop soil erosion	
4. Finches diversified in order to survive	
5. Germs mutate to become drug resistant	
6. Parasites multiply to infect the host	
7. The sun makes light so that plants can photosynthesize	
8. Water condenses to moisten the air	
9. Molecules fuse in order to create matter	
10. Earthquakes happen because tectonic plates must realign	
11. Geysers blow in order to discharge underground heat	
12. The earth has an ozone layer to protect it from UV light	
13. Flowers wilt because they get dehydrated	
14. Bread rises because it contains yeast	
15. People get the flu because they catch a virus	
16. Zebras have black stripes because they eat coal	
17. Gusts of wind occur because animals exhale together	
18. Clouds form because bits of cotton collect together	
19. Children wear gloves to keep their hands warm	
20. Teapots whistle to signal the water is boiling	
21. People buy vacuums because they suck up dirt	
22. Cars have horns to illuminate dark roads	
23. Eyelashes developed so that people can wear mascara	
24. Mothers kiss babies in order to scare them	

Table 1.2. Explanations for "why things happen." Which do you consider correct?

Explanations 1–12 in Table 1.2 were considered as scientifically unwarranted, teleological explanations. Explanations 13–24 were control ones designed to track participants' abilities to evaluate sentences at speed (13–15 and 19–21 were correct explanations; 16–18 and 22–24 were incorrect). Now you can see how you did! Overall, participants judged accurately most of sentences 13–24. But fast-speeded participants accepted, on average, more of the unwarranted teleological explanations than participants in the other groups. It seems that not having a lot of time to carefully consider a statement makes the expression of one's intuitive thinking more likely. But this was also, more or less, the case for at least half of those who had adequate time to think about the statements.[10]

Therefore, even though the tendency of adults to provide teleological explanations is less generalized compared to children, it emerges when they do not have enough time to think. Adults can be considered as more "expert knowers" than children, but what is the case for actual science experts? Does expertise in science affect the tendency to provide unwarranted teleological explanations? This was investigated in another study that involved scientists working in chemistry, geoscience, and physics, as well as two control groups of college undergraduates and people who were age-matched to the scientists but had only bachelor's degrees.[11] The test consisted of one hundred single-sentence explanations (thirty test explanations and seventy control explanations) for "why things happen," and participants were asked to describe the statements as "true" or "false." The test explanations were scientifically unwarranted teleological ones for natural phenomena, such as "Trees produce oxygen so that animals can breathe." Control explanations were included to track participants' biases, abilities to read at speed, and accuracy at judging explanatory statements in general. These control explanations were of four types: (i) true teleological explanations (e.g., "Women put on perfume in order to smell pleasant"); (ii) false teleological explanations (e.g., "Lamps shine brightly so that they can produce heat"); (iii) true causal explanations (e.g., "Soda fizzes because carbon dioxide gas is released"); and (iv) false causal explanations (e.g., "Oceans have waves because they contain a lot of seawater"). In each group, participants were randomly assigned to speeded and un-speeded conditions. The results showed that the physical scientists generally exhibited a lower acceptance of teleological explanations than participants in the two control groups. This suggests that decreasing acceptance of teleological explanations may be due to scientific expertise rather than to maturation. However, in all groups there was a higher acceptance of inaccurate teleological explanations under speeded conditions. This might mean that an intuitive bias in favor of teleological explanations persists, even if expert knowledge generally reduces their acceptance.

Enough with research findings—and these are just a few of the relevant published studies. What does all of this show? The main conclusion is that we tend to intuitively think in teleological terms, especially for the parts of organisms and the parts of artifacts, and this tendency becomes more evident when we respond spontaneously, without having a lot of time for reflection. How our brain works in this case is not entirely understood, but psychologists have made different proposals. According to one model, we have two complementary systems in our mind, System 1 and System 2. System 1 operates automatically, quickly, almost effortlessly, and involuntarily; System 2 deals with effortful mental activities, and its operations are associated with agency, choice, and concentration. Both systems operate when we are awake, with System 1 continuously generating suggestions for System 2 such as impressions, intuitions, intentions, and feelings. System 2 generally endorses the suggestions of System 1 and so most of what we think stems from our intuitive thinking. The problem with this is that, although System 1 is generally good, it also has biases and so it is error-prone under certain conditions. When System 1 runs into difficulties, it is System 2 that takes over and usually gets things right. However, because System 1 cannot be turned off and because System 2 has no clue of the biases, these cannot be always avoided, and so wrong intuitions can turn to beliefs.[12] This model could certainly explain the findings of the studies presented above and the persistence and intuitive-ness of teleological thinking.

Whether or not one accepts the particular model of System 1 and System 2, evidence strongly suggests that intuitions play an important role on our understanding of the world. In the present book, I follow the definition of intuition as "a judgment 1. that appears quickly in conscious-ness, 2. whose underlying reasons we are not fully aware of, and 3. is strong enough to act upon."[13] This means that spontaneously we come up with an explanation, and although we do not know exactly why this happens, we are ready to use it. But how does this work? When we are given insuf-ficient information about a topic, our brain completes the picture on the basis of our previous assumptions about the world.[14] It is in this sense that our early awareness of intentionality—that from a very young age we per-ceive people using objects for a purpose—may intuitively make us think about the parts of organisms in the same way we think about the parts of artifacts, that is, think of them in terms of intended uses, functions, or roles. Especially when we have no clues about what is going on, we intui-tively think in terms of intentions, goals, and design.

So, one thing to keep in mind is that based on our experience we unconsciously come up with explanations that seem to make sense. Many times they do, for example, when we step back from the edge of the roof of a tall building in order to refrain from falling. But in other cases, as in the explanations for the existence of parts of objects, these explana-tions may not be working well. The distinction between physical expla-nations and teleological explanations described in several of the studies above reflect two different stances that have been described as the phys-ical stance and the design stance, respectively. The physical stance is simply the use of whatever we know about physics (e.g., how objects fall to the ground) in order to make predictions or explanations. The physical stance generally works for all kinds of entities—organisms, artifacts, and nonliving natural objects (e.g., if I hold a plant, a watch, or an emerald and I suddenly release them, they will all fall to the ground). The design stance is a different strategy that relies on additional assumptions, which are that a specific object is designed and that it will operate according to that design. There is also a third one, the intentional stance, which can be considered as a subspecies of the design stance.[15]

Let me explain this last statement. I take the intention of the designer to be a property inherent in the design, because the properties of the designed object reflect the intentions of its designer. Think of a kitchen fork, which has multiple prongs (or tines). Can we claim that whatever has a prong is a kitchen fork? No, because pitchforks also have them. Does it make a difference whether the item has three or four tines, or more? No, because although kitchen forks usually have four tines, those for babies may have fewer. And although a pitchfork may look like a kitchen fork, we would not call it a "big kitchen fork." The reason is that we know that a kitchen fork is an object that we use in order to eat, whereas a pitchfork is an object that we use for pitching hay. In a similar manner, we can certainly distinguish between a knife and a sword, and we would never ask for a sword to cut the bread for dinner. How about chairs? We cannot define a chair as an object that has four legs, because tables also have four legs. There also exist kitchen chairs, office chairs, wheel chairs, arm chairs, and so on, and so there is no single way we can provide a general description of chairs based on their appearance. In this sense, it is the intended use of artifacts, what they were made for, that makes them distinct from one another.[16] This intention is inherent in the design of artifacts; thus the design of an artifact and the intentions of its designer cannot be really separated. Therefore, in this book I do not differentiate between the design and the intentional stances, and I refer only to the design one, which is broader.

The problem now is that whereas I can make similar predictions and explanations for a plant, a watch, and an emerald based on the physical stance, this is not the case for the design stance. If I release any of these objects from my hand from the top of a high building, they will all fall down. If all of these break, I can also explain this effect in terms of physics: the gravitational force brought the objects to the ground in an accelerated motion, and when they touched the ground a force (which I could actually estimate) was exercised on them, and this is why they broke into pieces. In contrast, the design stance does not allow me to see these objects in the same way. An emerald that broke into two pieces may have now become two smaller emeralds. However, a watch will not turn to two smaller watches when it breaks; it will only turn into a broken watch. Things are similar for the plant: if it falls to the ground and breaks into two pieces, it will be a broken plant. Now, the major issue—for the purposes of this chapter—is that the design stance makes us think of the plant in the same terms as the watch rather than as the emerald, even though both the emerald and the plant are natural objects and not artifacts. Therefore, a serious concern is to clarify how organisms, artifacts, and nonliving natural objects relate to one another.

As is evident from the example I used, and from the studies presented above, nonliving natural objects are rather easy to distinguish from the others. It is only the physical stance that works for them, and adults do not seem to think of them in terms of purposes, even though children were found to do so in some of the studies. But when it comes to organisms and artifacts, and especially their parts, things can become very tricky. Let us get back to one of the questions asked in the beginning of the chapter. Are emeralds and plants green because they consist of green parts or because being green helps there be more of them? The correct answer is without doubt that both emeralds and plants are green because they consist of green parts. Emeralds have green color because they contain traces of chromium and vanadium. Plants are green, or have parts that are green, because they contain chloroplasts that are small, intracellular organelles filled with chlorophyll. Therefore, the physical explanation "they are green because they consist of green parts" works perfectly for both emeralds and plants. Yet the physical stance is not enough for us! We cannot refrain from thinking at the same time that chloroplast and chlorophyll are involved in photosynthesis, the process through which plants absorb solar energy and build organic molecules such as glucose (which both themselves and animals, including us, then consume). If plants were not green, that is, did not have chloroplasts, they might not be able to absorb solar energy and built glucose. Then they might not be able to live—it is no coincidence that most plants have green parts such as leaves.

The issue, in other words, is that when it comes to organisms, physical explanations are appropriate but may look insufficient to us. The reason is that we think of organisms not only in terms of how they are or look, but also in terms of what they "do." Emeralds consist of various minerals that are ordered in a crystal structure, but they do not perform any function. In contrast, plants consist of cells that contain chloroplasts, which in turn interact with their intracellular environment and contribute to photosynthesis. Therefore, although both emeralds and plants are natural objects, there is something different about plants: they are alive. The constituents of plants are therefore not static, such as those of emeralds, but dynamic and involved in what we perceive as functions. This is where artifacts enter the scene. Artifacts are all of those objects that humans have designed *in order to* perform particular functions. They are not alive, of course, but they are functional. This is why we have always used artifact metaphors in order to better understand and describe organisms, the most predominant being the organism-as-machine metaphor. But this can also be problematic because it is possible for us to forget the differences between organisms and artifacts, and to focus on the similarities highlighted by the metaphors. And this is when misunderstanding occurs because, despite superficial similarities, organisms and artifacts are very different. Let us see why.

Earlier, I wrote that one might reasonably think that both the wings of airplanes and the wings of birds are for flying. If I asked you, "Why do birds have wings?" you would most likely reply, "In order to fly." (I have never received any other spontaneous response to this question in my courses and talks.) But if I started being more specific, things could become tricky. If I asked you, "Why do eagles have wings?" your response would most likely be, "In order to fly, of course," and perhaps you would also start wondering whether I am wasting your time with such questions. But if I asked you, "Why do penguins have wings?" what would you say? Penguins do not fly; they have relatively small wings for their body size, so it is actually impossible for them to take off and fly in the way eagles do. But then you might think, "Well, penguins use their wings in order to swim, and they are actually able to swim very fast underwater". So, you could eventually respond, "Penguins have wings in order to swim." Sure, you *could*. But what if I asked you, "Why do ostriches have wings?" Ostriches use their wings neither to fly nor to swim. What would you reply in this case? Apparently, the intuitive response "In order to fly" to the question "Why do birds have wings?" does not work well. All birds have wings, but not all of them use their wings in order to fly. Birds, like all organisms, have come to possess their features through evolution and are not intelligently designed. All birds have wings—which are actually their forelimbs —but not all of birds use these forelimbs for flying.

Let us now consider artifacts. What if I asked you, "Why do airplanes have wings?" A reasonable response would be, "In order to fly." Indeed! All airplanes have wings in order to fly; and there are no airplanes without wings. There exist other aircrafts, of course, that do not have wings, such as helicopters. But there are no airplanes without wings. Furthermore, airplanes have wings that are proportionally large to their size. A Cessna airplane has smaller wings than an Airbus, and in both cases the wings are long enough to allow takeoff and flight. There is no way that you would find an Airbus with the wings of a Cessna, or vice versa. Why is this the case? Because airplanes are artifacts and they have been intentionally designed for a purpose: to fly. Therefore, they have wings that are appropriate for this purpose. There is no way that you would find an airplane that uses its wings to swim (a submarine with wings?), or another that would have wings too small or large to preclude it from takeoff. You may find airplanes that do not fly, for instance, in museums, but this use (sitting in museums) is not what they were designed for—this is just an incidental use. Usually, airplanes as all artifacts function properly, but of course things can go wrong, either because of bad design or because of bad maintenance or use. But in all cases, the wings of airplanes are there in order to allow the airplane to fly. This is why only airplanes have wings; you will not find wings on ships or trains! Artifact parts may have incidental uses, but more often than not they reflect their intended use. The "in order to fly" response would work for any airplane because airplanes are designed in order for people to travel around the world in the air. The wings of airplanes are designed in order to allow takeoff, flight, and landing. But this is not the case for birds.

"OK then," you might think. "I will not say that birds have wings for flying, but I will say that eagles have wings for flying, penguins have wings for swimming, and ostriches have wings for . . . nothing obvious, but they should be doing something. Is there a problem with that?" Unfortunately, yes, if you like to think rationally. There is a problem with this kind of answer. The idea that everything must exist or must be the way it is for a purpose is unwarranted unless we can empirically establish that this is the case. In the case of artifacts, things are simple. Artifacts are intentionally designed by humans for a purpose. Therefore, we only have to ask its designer or look into the operating manual to find out the function of an artifact and its parts. But when it comes to organisms, things are very different. Organism parts do not have intended uses, only evolved ones. As a result, a lot more of the uses are incidental—wings can be used for flying, for swimming, or for nothing at all. To see what exactly is the case, one cannot ask the designer or look at the operating manual, but only study the life of organisms in their natural environment. Here is where the conflict between our Systems 1 and 2 might arise. System 1 suggests that the wings of birds exist in order for them to fly. Given, perhaps, our experience with airplanes, System 2 easily accepts this explanation, and thus an intuition turns to a belief. Only when we encounter difficulties, for instance, by considering ostriches and not just eagles, can System 2 become more activated, take over, and produce the reasoning I have outlined above. But, naturally, we tend to rely on our intuitions to explain phenomena. Therefore, when it comes to understanding organisms, scientific explanations can thus be counterintuitive, or at least less intuitive than the explanations we can derive from our knowledge and understanding of artifacts.

We can think of the wings of birds in a similar way as the wings of airplanes in order to describe and explain flight. But we should think of the wings of birds that fly as *if they were* something similar to the wings of airplanes, and not that the wings of birds *are indeed* similar to the wings of airplanes. Such metaphors have a limited explanatory use; we should not take metaphors literally. Design is literal for artifacts, but it can be used only in a metaphorical sense for organisms. One might accept that in the case of organisms, "the 'designer'—in a strictly metaphorical sense—is the unconscious process of natural selection, the blind watchmaker," as Richard Dawkins put it.[17] But there is actually no watchmaker, blind or not. Life simply goes on; in the short term, individual organisms develop, live, and die. In the long term, species evolve. In all cases, unconscious natural processes and not conscious agents are involved. Therefore, I do not see the point of talking about natural design in the case of organisms, even in the metaphorical sense. This does not mean that there are no natural functions, for example, that we cannot say that we have hearts in order to pump blood. Indeed, we can. But these are evolved functions, not designed ones. When something exists because it confers an advantage to its bearers, who in turn survive and reproduce better than others, it is legitimate to say that this something exists for doing whatever it does because it has been *selected for* doing so. But there is no design here; this is natural selection. The problem that hinders our understanding of organisms is design and design-teleology, not teleology in general.[18]

The above points to an important distinction. We often think of the entities around us as "living" and "nonliving," and this is how we often learn at school to classify them. However, the distinction between organisms and everything else that is not alive blinds us when it comes to design. Therefore, I suggest that a better distinction to make is that between human-made objects and natural objects—in other words, between arti-facts and everything else. The latter category will include organisms and nonliving natural objects. We can thus imagine a continuum of entities. At one extreme lie artifacts, such as watches and airplanes, which are the nonliving products of humans, intentionally designed for a purpose; at the other extreme lie natural objects, such as emeralds and clouds, which are the nonliving products of nature and which are not designed. Organisms, including humans, are products of nature like natural objects and exhibit design like artifacts; therefore, they could be considered as being somewhere between the other two. But the big question is, where exactly?[19] This answer notwithstanding, we should be careful to clearly differentiate between organisms and either nonliving natural objects or artifacts. Whereas explicitly we seem to succeed in both cases, we may fail to do so implicitly and apply artifact thinking to organisms. It is often easy to distinguish between an organism and a rock, and explain their dif-ferences in terms of exhibiting or not certain properties that we usually associate with the state of being alive (e.g., reproduction, development, sensitivity, etc.). However, in many cases it is a lot more difficult to dis-tinguish between the functional parts of organisms and artifacts, such as wings. The design stance certainly has had and still has a heuristic value, because through that we can make sense of nature in particular cases. But the conclusions about organisms to which we might arrive based on the design stance may be wrong.

In the next three chapters I describe what many people tend to think about human development, human life, and human evolution. The notions of genetic fatalism, destiny, and intelligent design seem to be prevalent in human thinking. My aim is not only to show what exactly people think but also to argue that these notions stem from our design-teleological intuitions. I suggest that these seemingly different notions—to which I will collectively refer as the "design stance"—stem from our robust, underlying, teleological intuitions, evidence for the existence of which was presented in this chapter. If you have found them in yourself, in the next chapters you will see what impact they may have on your thinking. But before doing so, a note of caution is necessary: not all intuitions are bad. My main argument is rather that some intuitions can be obstacles in understanding the concepts and the conclusions of science. Despite their well-documented misunderstandings of natural phenomena, young children are at the same time capable of identifying statistical and causal patterns, which in turn are essential for understanding natural phenomena. The ability of scientists to detect correlations, to infer causal connections for these, and eventually to reveal the mechanisms that explain the respective phenomena emerge in very young ages.[20] Therefore, whereas some intuitions are really useful, in *Turning Points* I only discuss the design stance that hinders our understanding of the living word.

Chapter 2

"OUR FATE IS IN OUR GENES"

J ames D. Watson is a famous Nobel Prize laureate who co-discovered the structure of DNA back in 1953. He was also the first director of the Human Genome Project, which aimed at "reading" the sequence of all human DNA. Watson famously stated back in 1989, "We used to think that our fate was in the stars. Now we know in large measure, our fate is in our genes."[1] Walter Gilbert, the co-developer of a method for "reading" DNA, and also a Nobel Prize laureate, was also a big proponent of the Human Genome Project: "Three billion bases of sequence can be put on a single compact disk (CD), and one will be able to pull a CD out of one's pocket and say, 'Here is a human being; it's me!' But this will be difficult for humans. . . . To recognize that we are determined, in a certain sense, by a finite collection of information that is knowable will change our view of ourselves."[2] Watson and Gilbert, two prominent scientists, referred to *fate* and *determinism*; what do these terms imply? *Fate* is the view that there is an underlying plan that determines how life events will turn out to be, which is outside our own control. *Determinism* is the idea that there exist certain factors that cause events to occur in particular ways. Both Watson and Gilbert expressed the view that genes hold the key to our lives, by possessing the information of who, what, and how we are. It was the late 1980s and early 1990s when Watson and Gilbert made these comments, and arguments like these—as well as the prospect of curing human genetic disease—supported the funding and the initiation of the Human Genome Project.

The underlying idea behind statements like these is genetic determinism: the idea that genes invariably determine the features of organisms independently of the environment in which these organisms live. This entails that when one has specific genes, it is possible to predict what kinds of traits, including diseases, one will have, without the need for any additional consideration of the environment in which that individual developed. This is an implicit message in the statements of Watson and Gilbert. Gilbert was also quoted stating that "sequencing the human genome is like pursuing the holy grail," whereas Watson also wrote that "the Human Genome Project is much more than a vast roll call of As, Ts, Gs, and Cs: it is as precious a body of knowledge as humankind will ever acquire, with a potential to speak to our most basic philosophical questions about human nature, for purposes of good and mischief alike."[3] Reading the human genome was therefore considered like reading the holy grail of life, which would therefore enable us to answer the most fundamental questions about what it means to be alive. To use a naïve metaphor, these statements convey the message that to understand how a delicious meal was made you only need to read the recipe. As if nothing else matters (I return to this point below).

When the completion of the sequence of the human genome was publicly announced in June 2000, optimism had not declined. This was described as "a milestone for humanity."[4] President Bill Clinton famously proclaimed: "Today we are learning the language in which God created life."[5] Polls soon followed that aimed at documenting the understanding of genes among Americans. For instance, the Opinion Dynamics Poll for Fox News, conducted in June 2000, asked the following question: "Earlier this week scientists announced they have completed a genetic map of the human body. If knowing your genetic code could tell you whether you would contract an incurable disease or not, would you want to know if you had that gene?" Whereas 7 percent of participants were not sure, 59 percent answered yes and 34 percent answered no. Another question was, "Do you think the ability to genetically map the human body is more likely to be a positive scientific discovery that reduces sickness and suffering, or a negative discovery that leads to lawsuits and ethical conflict?" This time, 16 percent of participants were not sure; but 56 percent thought of the discovery as positive, and only 28 percent thought of it as negative. Another poll, conducted for NBC News and the *Wall Street Journal* prior to the announcement of the sequencing of the human genome, asked the following question: "Later this month, scientists are expected to announce that they have completed a genetic blueprint of the human body. Some are hopeful that this will help detect and combat illnesses. Others are concerned that this could violate privacy rights because information about people's health problems may be used against them. Does this concern you, or not?" More than half of the respondents, 56 percent, replied that they were concerned, whereas 42 percent replied negatively.[6] Notwithstanding the differences between the two polls, which ask the same privacy question differently, these results showed that more than half Americans would like to know if they had a gene related to an incurable disease, that they believed that genetics discoveries might reduce sickness and suffering, as well as that such information about health might be used against them. The underlying idea in these views is that if one has the gene, one will also develop the disease and therefore the relevant genetic information might be used for fighting the disease as well as against that person.

More interesting are the results of polls when it comes to human behavior. Based on a Gallup poll conducted in 2014 in the United States, 42 percent of Americans answered that people are born gay, whereas 37 percent answered that people become gay because of factors such as their upbringing and environment. Interestingly, and despite the differences in the wording of the question (the term "homosexuality" was used between 1977 and 2008; the terms "gay or lesbian" were used in the more recent version), there has been an increase in the view of homosexuality as innate rather than as acquired during the last forty years. In 1977, for instance, only 13 percent of respondents believed that homosexuality is something a person is born with, whereas 56 percent believed that it is due to upbringing and environment.[7] Another poll, conducted in 2013 by Pew Research Center, found that 42 percent of Americans answered that being gay or lesbian is "just the way some choose to live," while 41 percent answered that "people are born gay or lesbian," and 8 percent said that people are gay or lesbian due to their upbringing.[8] Views like these are not restricted to the United States. For instance, polls in Canada and Great Britain, conducted in 2004, showed that 54 percent and 55 percent, respectively, believed that homosexuality is "something a person is born with."[9] In contrast, 29 percent of Canadians and 24 percent of British adults answered that it is upbringing and environment that are behind homosexuality.

"Something a person is born with" is usually attributed to one's genes, although—as I explain in part 2—it should be attributed to one's development, which includes genes and a lot more. But genes are often perceived to contain the "recipe" of how we are when we are born, and so people usually think of innate characteristics as genetic, even though they are not all the time. Polls only provide some very rough results from which various inferences can be made. It is therefore more interesting to look at the academic research on this topic, which typically goes deeper into how people think. For instance, a study with sixty-three teenage American students investigated in some detail their understanding of genes and of gene function. One of the questions asked was straightforward and simple: "What are genes and what do they do in the body?" Here are some responses that students gave:

1. "genes carry our traits"
2. "genes are our features"
3. "genes control/determine our traits"
4. "genes decide who we are and what we will look like"
5. "genes tell us what we will look like"
6. "genes are like maps of our traits"
7. "genes carry information about our traits"

Please make a note of any of those responses with which you agree—a check mark next to each of them will do. Then please choose the one with which you agree most before you read below. Researchers considered statements such as (1) and (2) as indicating a view of genes as passive particles associated with traits, because there was no reference to what genes "do." Statements such as (3) and (4) were considered as indicating a view of genes as active particles because there was reference to what genes "do." Finally, responses such (5), (6), and (7) were considered as indicating a view of genes as containing instructions. In this study, fifteen students were found to think of genes as passive particles, thirty-three were found to think of genes as active particles, and twelve were found to think of genes as containing some information (three students did not give any response). The problem is that, except for statement (7) above, all the others are scientifically illegitimate in some way. Furthermore, no response described genes as containing information for the synthesis of molecules (RNA or proteins)—which is what genes "do."[10]

More than half of the students thought of genes as active particles that determine and control traits, but is this conception prevalent beyond the confines of this study? When other researchers analyzed 500 students' essays, submitted for a national contest in the United States, they found that about one out of ten of these essays (12.8 percent) included genetic determinist conceptions.[11] The key ideas in this case are that genes determine all traits, that one gene determines one trait, that there is lack of environmental influence, or that there is lack of multigene involvement in traits. At first sight, 12.8 percent does not look a lot. However, one should consider that the essay contest did not have the aim of eliciting students' conceptions, and it only included general questions, such as "If you could be a human genetics researcher, what would you study and why?" or "In what ways will knowledge of genetics and genomics make changes to health and healthcare in the United States possible?" Responses can be different when the questions asked are more explicit. For instance, when students of a similar age in Great Britain provided answers to the question "Why are genes important?" 73 percent of them responded that genes are involved in the determination of characteristics, and only 14 percent said that they are involved in the transfer of information.[12]

But what exactly could "determination of characteristics" be about? Our tendency to accept that genes are the major factors that shape traits becomes more evident when it comes to human behavior and disease. A quick search on the World Wide Web reveals several examples. For example, a 2014 article in the *Guardian* was titled "'Happy Gene' May Increase Chances of Romantic Relationships."[13] According to that article, scientists claimed to have found a gene that increases the chances that university students are in a romantic relationship. The scientists focused on a gene called *5-HTA1*, which affects the levels of the hormone serotonin in the body. One version of the gene, the C variant, leads to higher levels of serotonin than the G variant. Tests on 579 Han Chinese students showed that half of those who had inherited two copies of the C variant, one from each of their parents, were in relationships. But students who carried one or two copies of the G variant had only a 40 percent chance of being in a relationship. Think about this. If you are indeed in a romantic relationship, it could be because of your genes, according to this article. So, if you have two copies of the C variant you have a 50 percent chance of being in a romantic relationship—but also a 50 percent chance not to be! In contrast, if you have one or two copies of the G variant, which according to the article makes people who carry it "more likely to be neurotic and develop depression," your chances of being in a relationship are 40 percent. This 10 percent difference was considered to be small but still significant.

Let us consider another example of how the alleged impact of genes on human behavior is reported. The title of a 2015 article in the *New York Times* suggested that "Infidelity Lurks in Your Genes."[14] In a study of about 7,400 Finnish twins who had all been in a relationship for at least one year, researchers looked at the link between promiscuity and specific variants of vasopressin genes, a hormone that has powerful effects on social behaviors like trust, empathy, and sexual bonding in humans and other animals. The study found a significant association between five different variants of the vasopressin gene and infidelity in women only. Actually, 40 percent of the variation in promiscuous behavior in women could be attributed to genes, according to the article. This means that 40 percent of the differences observed for this particular behavior was due to differences in genes. So, if you are a woman who has cheated or is thinking of cheating on your partner, it may not be your or your partner's fault. It may not be that you are unhappy with this relationship, for whatever reason; it may not be that you are taking your partner for granted and do not appreciate what your partner does for you; it may just be that infidelity is in your genes, and that you cannot control it. Although the author of the article acknowledged that single genes rarely determine behavior and that correlation does not entail causation, the final conclusion was that much as people may not want to cheat on their partners, "sexual monogamy is an uphill battle against their own biology." So, if you are a man who is not in a committed relationship, next time you meet a woman you like, you might arrange the first date at a genetics laboratory to find out if she carries the variants of the vasopressin gene related to infidelity. It could be a waste of time to try to build a relationship with a woman who, as the author of the article put it, might have "a propensity for sexual exploration that seemed in some ways independent of the emotional status of her relationships." (I am of course being sarcastic here.)

But do people really think in such ways? A study that aimed at exploring how the public interprets articles on the genetics of behavior gave American participants one of three published news articles to read: "'Liberal Gene' Discovered by Scientists"; "Born into Debt: Gene Linked to Credit-Card Balances"; and "Key Breast Cancer Gene Discovered" (the latter were considered the control group of the study, as the link between cancer and genetics is well-established).[15] Participants were asked to mention how much they believed that fourteen different characteristics were "influenced by an individual's genes as opposed to his or her environment and choices." They then had to choose one of the options listed in table 2.1 below. Why don't you try this yourself? Like the participants did, imagine that there is no right or wrong answer; just choose what you think by putting checkmarks in the appropriate boxes.

Characteristic	% of Characteristic Being Genetic										
	0	10	20	30	40	50	60	70	80	90	100
Skin color											
Natural hairstyle											
Height											
Breast cancer											
Intelligence											
Sexual orientation											
Obesity											
Mathematical abilities											
Alcoholism											
Behaving violently											
Gambling addiction											
Being liberal or conservative											
Having credit card debt											
Preferring Apple or Microsoft											

Table 2.1. Do genes affect the expression of these characteristics? If so, to what degree (in percent)?

The results showed that the more biological the characteristic (e.g., skin color, height), the more genetic it was considered. When the biological nature of a characteristic was less clear, genetics was perceived as playing a less important role. Thus, those participants who read the article on the "liberal gene" considered that the genetic attribution for being liberal or conservative was significantly higher compared to the control group. Similar was the case for sexual orientation, mathematical abilities, alcoholism, violent behavior, gambling addiction, and having credit card debt.

At the same time, those participants who read the "debt gene" article, attributed to genetics—at a rate significantly higher than the control group—not only having credit card debt but also all of the other characteristics (except for skin color, natural hair style, height, and breast cancer, for which many participants in both groups made genetics attributions). Results like these show that people can draw mistaken conclusions from what they read in the news. What also becomes evident is that the idea that a gene determines, or simply affects, a human characteristic is quite intuitive and can be elicited and expressed even by a single influence such as reading a news article about the supposed power of genes.

A big problem is that we cannot really tell how "genetic" or "environmental" a characteristic is. It is impossible to measure the relative contributions of genes and environment to the development of our characteristics. The reason for this is that these contributions are not distinct and additive, but interdependent. To use a classic example, imagine two workers who lay bricks to build a wall. In this case, it is quite easy to measure the contribution of each of them to building the wall. If one laid forty bricks and the other laid sixty bricks to build the wall, we can say that the building of the wall was 40 percent due to the former and 60 percent due to the latter. But this can be estimated because the contribution of each worker is independent—each one of them lays bricks on the wall independently. However, if one of the workers mixes the mortar and the other lays the bricks, it is impossible to measure their relative contributions to the overall completion of the wall. The reason for this is that these contributions are interdependent and cannot be distinguished from each other. Even if we could measure the quantity of mortar and bricks in the wall, this would not be a measure of each worker's contribution, as both mortar and bricks are needed for building the wall and cannot be considered separately.[16] It is in a similar manner that the contributions of genes and environment are interdependent and thus cannot be considered separately. Therefore, the perception that some characteristics are more or less genetic than others is inherently wrong.

In 2013, the *New York Times* published an essay by actress Angelina Jolie in which she revealed that she had undergone a double mastectomy because she had been found to carry a "faulty gene," *BRCA1*, that "sharply" increased her "risk of developing breast cancer and ovarian cancer."[17] Given that her mother had died at the age of fifty-six after battling cancer for ten years, Jolie decided that she could eliminate the risk of developing cancer by removing the organs in which it could develop. Jolie concluded her essay by writing: "I choose not to keep my story private because there are many women who do not know that they might be living under the shadow of cancer. It is my hope that they, too, will be able to get gene tested, and that if they have a high risk they, too, will know that they have strong options. Life comes with many challenges. The ones that should not scare us are the ones we can take on and take control of." Since then, headlines relevant to this story have appeared in the news: "Women Like Angelina Jolie Who Carry the *BRCA1* Gene Are Less Likely to Die from Breast Cancer if They Have Their Ovaries Removed"; "Study: Women with *BRCA1* Mutations Should Remove Ovaries by 35"; "Moms with *BRCA* Breast Cancer Gene Mutations Face Tough Decisions."[18] What is the problem here? Particular versions of the *BRCA1* and *BRCA2* genes have been associated with breast and ovarian cancer. These genes produce proteins that help repair damaged DNA and thus contribute to the stability of the genetic material of a cell. When certain genetic changes (mutations) occur, DNA damage may not be repaired properly and so cells are more likely to acquire additional mutations that may contribute to the development of cancer (more on this in chapter 7).

To assess the impact of the Angelina Jolie story, a survey was conducted in the United States.[19] Participants were asked to describe their sources of information for this story, as well as their understanding of and reaction to it. They were also asked some hypothetical questions. Below are some of them. Before you look at the results, why don't you try to see what answers you would give?

#3. Angelina Jolie carries a "faulty gene" associated with breast cancer. Because of this gene and her family history, what was her lifetime risk of developing breast cancer? Even if you are unsure, please provide your best estimate. [RANGE: 0–100]

_____ % chance of developing breast cancer

#5. To the best of your knowledge, the faulty breast cancer gene is responsible for what percentage of all cases of breast cancer? Even if you are unsure, please provide your best estimate.

01 Less than 5%
02 Between 5% and 15%
03 Between 16% and 35%
04 Between 36% and 50%
05 Between 51% and 75%
06 Between 76% and 100%

#6. To the best of your knowledge, what is the average woman's risk of developing breast cancer over her lifetime if she does not have the faulty breast cancer gene? Even if you are unsure, please provide you best estimate.

01 Less than 5%
02 Between 5% and 15%
03 Between 16% and 35%
04 Between 36% and 50%
05 Between 51% and 75%
06 Between 76% and 100%

Among the participants who were aware of the story, about half mentioned a risk that was within a reasonable range (80–90 percent) of Jolie's reported estimated risk of 87% in their responses to question 3. Nevertheless, only about one in five of these participants correctly stated the contribution of *BRCA* mutations to all breast cancer cases (the correct response to question 5 was considered by the researchers as less than 5 percent), whereas about a third of them were aware of an average woman's risk of getting breast cancer over her lifetime if she didn't have a *BRCA* mutation (the correct response to question 6 were considered by the researchers those between 5 and 15 percent). Finally, only 8.9 percent of these participants answered correctly both questions and therefore had the information that was necessary in order to understand Jolie's risk of developing cancer. This means that whereas three quarters of the participants were aware of the story, approximately only one out of ten had a correct understanding of the issues at stake. Why? Apparently, many people thought that the impact of *BRCA* genes in cancer is larger than what it actually is.

Once again, we see evidence that people tend to exaggerate the impact of genes on disease. Whereas there is certainly a relation between breast cancer and the *BRCA* genes, this is a probabilistic relation (in fact, a statistical correlation). This means that someone who has these genes also has a higher probability to develop the disease than those who do not have them. Having a higher probability means that among those who have the gene, more developed cancer compared to those who do not have it. However, this does not mean that all of those who have these genes will definitely develop the disease. There are of course cases, like that of Angelina Jolie, in which this probability is very high. However, most cases of breast cancer are not due to the *BRCA* genes; in fact, the mutation that Jolie was found to carry is relatively rare. But this is not the message that people in this study got from her story. This brings us to a distinction not often made in the public sphere, between a genetic and a hereditary disease. Hereditary diseases, that is, those that run in families, are usually genetic as well, because of genes that are passed on from one generation to the next. However, there exist genetic diseases that are not hereditary, and cancers are an exemplar case. Cancers are due to mutations in our DNA, and therefore are genetic; however, more often than not these mutations occur during one's life and are not inherited from one's parents.[20] This is why most cancers are not hereditary.

It is also interesting to look at attributions made to genes in an implicit manner. In one study, forty-seven American adults were asked to place terms into one of two opposing categories.[21] First they were asked to put "gene" terms (e.g., *genome, DNA, heredity*) and "socialization" terms (e.g., *learn, training, experience*) into their respective categories. After that, they also did the same with "fate" terms (e.g., *plan, necessity, destiny*) and "choice" terms (e.g., *free-will, decision, selection*). Then, participants were asked to put "fate" and "gene" terms into one category, while putting "choice" and "socialization" terms in another category, as quickly as possible. They were then asked to do exactly the opposite, that is, to put "fate" and "socialization" terms into the same category, while putting "choice" and "gene" words into another category. Although you cannot take the test here, you can try a much simplified version: which of the terms below would you associate with "genome," "DNA," "heredity," and "bloodline"? You can try to fill in table 2.2 below, placing each term in the appropriate category.

Terms	Associated with "genome," "DNA," "heredity," "bloodline"	NOT associated with "genome," "DNA," "heredity," "bloodline"
blueprint, option, destiny, selection, certainty, free-will, god, decision, plan, opinion, necessity, permanence, freedom, preference		

Table 2.2. Why don't you try to see which associations you find legitimate? No time constraints in this version!

Participants were slower in putting gene and choice terms in the same category while also putting socialization and fate terms in the same category, compared to putting gene and fate terms in the same category and socialization and choice terms in the same category. This suggests that participants implicitly associated genes with fate and socialization with choice, rather than the opposite. This indicates that people implicitly think of genes as determinants—in a broad sense—of conditions and phenotypes.

That people tend to implicitly think of genes as determinants was also the finding of another set of studies in Canada, which aimed at investigating what consequences people draw from a perceived genetic cause for obesity. In the first study, undergraduate students were asked to indicate whether they believed that obese people can control their weight, as well as whether obesity is mostly due to genetic or environmental factors. The results suggested an association between a belief in a genetic basis for obesity and a belief that obese people cannot control their weight. This association was further explored in another study, in which undergraduate students were asked to express their beliefs about a phenomenon related to obesity (metabolic rate), given particular explanations. In this case, researchers manipulated the explanations and evaluated their effects on people's beliefs about weight. The results indicated that a genetic influence for high metabolic rate was considered as more important than an experiential influence. Finally, in a third study, undergraduate students read one of three fictional media reports presenting a genetic explanation, a psychosocial explanation, and no explanation for obesity, respectively. Then, they were asked to participate in a food-tasting task in order to evaluate the flavor of cookies. Participants sat in front of a large bowl of broken chocolate chip cookies, and the experimenter left the room, instructing them to taste and evaluate the cookies. The bowl was secretly weighed before and after the participants ate the cookies. The difference in weight was used as an indicator of cookie consumption, and the researchers found that participants who had read the genetic explanation ate significantly more cookies than the others.[22] From results like these one can infer the perceived power of genes. Those who read the genetic explanation might have concluded that there was nothing they could do to control their weight anyway, and so they ate more cookies than the others did.

These different research findings all show that people think that genes determine biological characteristics and behaviors. Participants in the above studies thought that there exist "genes for" biological characteristics such as eye and skin color, behaviors such as debt and obesity, and diseases such as cancer. The common underlying idea in most of this research seems to be that these characteristics are the expression of an inherent potential somehow registered in our genes. But where is the design stance in all of this? There are two distinct conceptual design aspects when it comes to genes. The first one is that genes could be perceived to exist for a purpose, that is, in order to determine a characteristic such as eye color. This idea may be enhanced by the shorthand "gene for" something, in which case people may think that the so-called "gene for" blue eyes may indeed exist for the purpose of developing blue eyes. The second one is that genes are perceived to do whatever they do invariably, as if the final outcome is somehow prescribed within them. In the latter sense, the design is inherent in the idea that genes contain the information for characteristics (genes as blueprints), so that one can think that the presence of the gene guarantees the emergence of the characteristic, as well as that the presence of the gene can be inferred by simply observing the characteristic. Both of these ideas, "genes for" and "genes as blueprints" are problematic for understanding genes, as they could enhance fatalistic views, such as those expressed, sometimes explicitly and other times implicitly, by participants in the research presented in this chapter.

The views of Watson and Gilbert, mentioned at the opening of this chapter, are no longer shared by most scientists, as I explain in some detail in part 2 of the book, especially in chapters 6 and 7. However, as the research presented above shows, the contemporary scientific understanding of genes and genetics/genomics research has not yet reached the general public. In contrast, people seem to think of genes in unwarranted ways, with important implications for their understanding of life, health, and identity. A complicated picture has been revealed by the findings of the Human Genome Project and several other subsequent studies, such as the ENCODE (Encyclopedia of DNA Elements) Project and various genome-wide association studies (GWAS). This picture can be summarized in the following statements: Genes are DNA segments that encode functional products but do not invariably determine our traits (including our characteristics and disease). Genes actually do nothing on their own, but they are implicated in the development of traits and disease, and they also account for variation in traits in particular populations.[23] This means that genes do not have much to do with fate, although they certainly make important contributions to our lives. The important point is that the relation between genes and traits is a many-to-many relation. This means that many genes are implicated in the development of (but do not *determine*) a single trait, as well as that a single gene may be implicated in the development of (but, again, does not *determine*) several different traits. This is why people with the same version of the same gene may have very different characteristics, either because of several other genes involved or because of other phenomena that took place during development (some examples are discussed in some detail in part 2 of the present book).

Here is a simpler way to conceptualize the role of genes in development (a naïve metaphor that I have already mentioned above, which has limitations like all metaphors). If you think of genes and DNA as a recipe including a generative plan, that is, a plan for how to make a certain meal, and if you think of different individual organisms as developmental systems that might implement the same instructions in a slightly different manner like two chefs might slightly differ in implementing the same recipe, then you can imagine different outcomes emerging. If you asked a famous chef to prepare dinner, she might be able to implement the recipe in detail and even prepare a meal that might look exactly like the one in the photo of the recipe book. If you asked me, you would most likely end up searching for takeout food, as my lack of patience and motivation for cooking would probably result to something you had better refrain from eating. This illustration shows that no matter what is included in the instructions, the final outcome depends not only on them but also on the developmental processes and on what happens during the implementation of those processes. This is why women have been found to carry genes associated with breast cancer without ever developing cancer themselves; this is also why women have developed breast cancer without carrying any such genes. Giving the recipe to me does not guarantee a nice dinner, because I must implement it under certain conditions; in the same sense, reading the DNA sequence and identifying the presence of a "cancer" gene does not guarantee cancer, because that gene must be expressed under certain conditions. In both cases, having the information is not enough.

The research presented in this chapter supports the conclusion that people tend to think in fatalistic terms when it comes to how we enter this world. The design stance may make us think of genes as somehow predetermining the outcome of our development, our characteristics, and our diseases. However, the conclusions from genetics research are that genes are not our essences, do not alone determine anything, and cannot alone explain why we have the characteristics we have. The if-you-have-the-gene-then-you-also-have-the-characteristic view is wrong, because development is a complex process. Our characteristics, and whether or not we will develop a certain disease, for instance, depend on unique interactions between our genome—which includes a lot more than our genes—and the environment(s) in which we live. Therefore, there is no genetic fate; and how we will be, look, or live is far from predetermined in our genes. Of course, genes impose certain limitations (e.g., no human will ever be born with wings), but beyond these the variation is enormous and the possible outcomes numerous.

In part 2 of *Turning Points*, I explain why there was no fate related to the existence of each one of us, neither regarding how we are and look, nor regarding our being here at all. Whether or not we are conceived, are born, and develop particular characteristics or diseases all depend on the contingencies of human reproductive and developmental processes. Particular turning points, as I show, can drive reproduction and development to one of several possible directions and outcomes. There seems to be no fate prescribed in our genes; or we should at least come to appreciate the importance of several critical events during our development, which are turning points.

Chapter 3

"EVERYTHING HAPPENS FOR A REASON"

Many people have the tendency to think that whatever happens in life, it happens for a particular reason and because it had to happen. People actually tend to interpret the outcomes of significant events in terms of design, that is, as if there was an underlying plan or intention in these outcomes. Such events, with a huge psychological impact all over the world, took place on September 11, 2001, in the United States. I remember watching the afternoon news on TV back in Greece when these attacks took place. Initially I thought that this was probably an accident with a small aircraft—a helicopter perhaps—which some stupid human mistake brought it upon one of the Twin Towers. Then, we got live images and I saw smoke coming out from one of the towers. Shock was affecting my ability to process what was happening; I could not fully realize what was going on. Suddenly, I saw a plane moving toward the towers on screen. I recall the presenter of the news saying that this was a video of the accident. I thought, "Wow . . . how could this terrible thing happen? A passenger airplane?" Then, I noticed that I was watching the airplane moving toward one of the towers, *while the other one was already on fire*. And it crashed. It was only then that I shockingly realized that I had just watched a *second* plane crashing into the second of the Twin Towers. The rest is history. Thousands of people died on that day, which has left a permanent mark in the memory of many people. The most significantly affected were of course those who died, and their families and friends. But there were also survivors, people who happened to be late to work on that day or who had happened to take a cigarette break and so were not in their offices in the towers at the moment of the attacks. A flight attendant was not on one of the planes because of scheduling errors. "God has a plan for you," people told her again and again. "You were meant to be here."[1]

This kind of thinking is very tempting. Back in my first year as an undergraduate student some twenty-five years ago, I remember talking to one of my classmates. She was from Cyprus, and she was born one day after I was born. Her father was a soldier in the Cypriot army, and he was killed during the Turkish invasion of July 1974; as I described in the preface, my father was an officer in the Cypriot army and was severely injured during the same events. Both of their wives were pregnant; my classmate and I were born in December of that year. She never met her father, whereas I was lucky to have my own father. I was always puzzled by the question: why was I lucky and she was not? My father still believes that there was a purpose; he was meant to be there for me, and to have another child—my brother—who would have not been born otherwise. It seems that he and many other people find reassuring the idea that there are no accidents in life and that things happen for a reason. But why is this reassuring? Perhaps because even terrible events might lead to a rewarding outcome—"every obstacle brings good," as my farther still says. This kind of thinking can be useful, as it may help people deal with and overcome problems and tragedies. If people think that something good might come out of a tragedy, they may find the courage to move on. But this kind of thinking can also have bad consequences, as we may perceive the good events that happen to some people and the bad events that happen to other people as natural. It can make us think that some people were meant to suffer whereas others were not.

Polls reveal the tendency of people to think like this. For instance, a 2005 Gallup poll in the United States, United Kingdom, and Canada, showed that about one out of four believed in astrology, that is, they believed that the positions of the stars and planets can affect people's lives. More women than men were found to hold such a belief.[2] Astrology is perhaps the best example of an activity that illustrates the human tendency to look for certainty in predicting future outcomes. The origins of astrology are now well understood, and its contribution to the origins of modern astronomy is well established. Serious astrology focused on the study of natural phenomena, particularly the weather, natural events, and human life and health. Celestial influences were thought to imprint a child with certain properties at the moment of its birth, which would in turn have an impact on its life. For instance, a child imprinted by a cold and dry planet such as Saturn might tend toward melancholy or depression, according to the tenets of astrology. But the aim of astrologers was not to predict anything about the future; rather, their aim was to identify possible tendencies or inclinations so that some kind of action could be taken, with, for example, changes in lifestyle and behavior. Thus, astrology was initially a serious pursuit, and it is no coincidence that notable astronomers such as Tycho Brahe and Galileo Galilei, among others, were involved in astrological activities. The best known of these activities was the horoscope: this literally means "observation of the time," and it contains the calculations at exact times of the positions of all celestial bodies relative to the horizon. It was with activities like these that astrology brought about the development of new mathematical tools and astronomical models and tables, and thus paved the way for the science of astronomy.[3]

But this is far from today's astrology, which claims to make predictions about one's future. Today, daily horoscopes provide general accounts for one's life and aims. For their conclusions astrologers generally use the horoscope, which is a picture that shows the positions of astrological objects within twelve equally spaced sectors as seen from a particular place and time on Earth. If this place and time are those of one's birth, then the horoscope is called a "natal chart." Astrologers claim to make inferences about one's character from such charts. But astrological accounts are not necessarily personalized. If you open several popular magazines and newspapers, you will most likely find pages devoted to astrology and filled with accounts about the characteristics of people depending on their zodiac signs. To which sign one belongs generally depends on the day of one's birth. These in turn belong to one of four elements (fire, earth, air, water), and signs belonging to same element are supposedly more compatible. Thus, Aries, Leo, and Sagittarius belong to fire; Taurus, Virgo, and Capricorn belong to earth; Gemini, Libra, and Aquarius belong to air; and Cancer, Scorpio, and Pisces belong to water.

Check out, for instance, my horoscope for December 27, 2016 (if you ask me what my sign is, I am supposed to reply that it is Sagittarius): "As fiery as you are—and that's pretty darned hot—when you say you're ready to get the show on the road, as far as a work project goes, there's absolutely no doubt that it's all going to come together, and quickly, too. You're set to pull off something like that now, and there'll be no stopping you, either—just as long as you manage your time carefully and don't waste a single moment. You can do it. You just need to believe you can do it."[4] What was the message in my horoscope that day? If you believe in something and you diligently try to achieve it, you will make it in the end. Is this news? Would anyone disagree with that? No, most likely not. The main issue here is that statements like these fit in the lives of most people. And those in need of motivation and support of any kind will tend to interpret their horoscope in the way that mostly works for them. Our minds tend to favor what already fits well with our thinking or experience, and at the same time ignore what does not fit very well (this is described as *confirmation bias*).[5] An important detail that we miss here is that any account of personality based on the time and day of one's birth should account for more than one person. Assuming that currently approximately seven billion people live on Earth, we can calculate that on average approximately 19 million people have their birthday on the same day. And from those people born on the same day, if we think that 24 hours is equal to 86,400 seconds, we can conclude that, again on average, about 220 people were born on the very same second of the very same day of a calendar year (e.g., 17:05:06 on December 16). As you surely realize, it makes no sense to think that descriptions like the one above can accurately account for all 220 of these people.

In 1985, a study was published in the prestigious journal *Nature*. Its author aimed at providing a double-blind test of the thesis that "the positions of the 'planets' (all planets, the Sun and Moon, plus other objects defined by astrologers) at the moment of birth can be used to determine the subject's general personality traits and tendencies in temperament and behaviour, and to indicate the major issues which the subject is likely to encounter."[6] The study included both scientists and astrologers as advisors, in order to avoid complaints of misinterpretations of the latter's views. The experiment had two parts. In the first one, 177 participants provided information that astrologers used to construct their natal charts and interpret them. Then participants were each given three interpretations of their natal charts. One of them was their own, and the others were randomly selected from the group; but participants did not know which one was which. The hypothesis of scientists was that this choice was random and so one in three participants would correctly choose his or her own natal chart purely by chance; the hypothesis of astrologers was that at least half of the participants would correctly choose their own natal chart. In the second part of the study (116 participants), astrologers were given at random a natal chart of a participant, this participant's independent measure of personality traits, and those of two other participants. Astrologers were then asked to suggest which of the three measures matched the personality description derived from the natal chart; in other words, the astrologers were tasked with finding which personality measure corresponded with the natal chart. The hypothesis of scientists was that one in three astrologers would make the correct choice purely by chance, whereas the astrologers predicted a correct choice at least in half of the cases. To avoid any biases, a double-blind approach was used so that neither the astrologers nor the experimenter knew which chart or personality measure came from which participant. A control group was also used. The results from part 1 were consistent with the scientific hypothesis without necessarily ruling out the astrological one (because participants could choose their personality trait profile at a better than chance level). The results from part 2 showed that astrologers performed rather poorly, making correct choices at the chance level (1/3) and below their predictions (1/2). Overall, the findings were considered to refute the astrological hypothesis, as astrologers generally failed to confirm their predictions in a test that was approved by them.

Now the problem is that despite its apparent flaws astrology is a popular subject, with "specialists" appearing in TV shows and writing regular columns in newspapers and magazines. At the same time, formal education seems not to do much to address the issue. For example, a study in Canada with first-year undergraduates aimed at documenting their attitudes toward astrology, as well as trying to find out whether they could distinguish between that and astronomy. It was found that among arts and science students, more than two out of three paid some or a lot of attention to horoscopes; about half of students from both groups found their horoscopes somewhat accurate; and between one and two out of ten of them had once or twice based conscious decisions on their horoscope (with science students generally doing a bit better in most cases than arts students). Finally, about half of arts students somewhat subscribed to the principles of astrology (that one's character and destiny can be understood from the positions of the sun, planets, and stars) and thought of both astronomy and astrology as being science, whereas this was the case for approximately one out of three science students. Interestingly, even though the science students seemed to have done better overall, the differences were not very significant. Yet one would expect science students to have done better than nonscience students.[7]

A few years later the same researchers replicated the same study, again in Canada, using almost the same questionnaire with a different sample of undergraduates. Their conclusions were disappointing, as the differences this time between arts and science students were even less significant, compared to the older study above. Again, more than two out of three of arts and science students paid some or a lot of attention to horoscopes; about two out of three of them found horoscopes somewhat accurate; about one out of five of them had based conscious decision on horoscopes once or twice; and about half of them somewhat subscribed to astrology and thought of both astronomy and astrology as being science. The researchers concluded that these results should alert people about the effectiveness of science education and about the possible influence of media on this topic.[8]

The results of a twenty-year (1988–2008) survey of science knowledge and attitudes toward science with approximately ten thousand undergraduates in the United States formed the basis for similar conclusions.[9] Only about one in five undergraduates responded that astrology is "not at all" scientific, and several of them agreed with the statement that "the positions of the planets have an influence on the events of everyday life." In particular, a large majority of students (78 percent) considered astrology "very" or "sort of" scientific. Also, about half of science majors said that astrology is "not at all" scientific. The researchers noted that this result should be interpreted with caution as these students could have been aware of the historical relation between astrology and astronomy. The researchers also found a small but significant correlation between science literacy and the ability to identify astrology as "not at all scientific." However, over the twenty years, researchers found little improvement in terms of how astrology is perceived. These researchers explicitly stated their concern about the status of school education and the aim to promote science literacy. But I am wondering: how many teachers explicitly address astrology in their science classes? This is a major issue for science education that often focuses on the transmission of content knowledge that students have to learn, rather than thoughtfully addressing their conceptions of real-life issues such as belief in astrology.

Unsurprisingly, the situation is not different in Europe. In one study, researchers analyzed data from a survey involving approximately one thousand participants from twenty-five European Union member states.[10] A central objective was to find out how scientific participants considered astrology. To investigate this, participants were asked to indicate on a scale from 1 (not at all scientific) to 5 (very scientific) their perception of ten subjects, including physics, medicine, astronomy, economics, history, homeopathy, psychology, biology, mathematics, and either "astrology" (half of the participants) or "horoscopes" (the other half). The researchers found that many more Europeans considered astrology as more scientific than horoscopes. In particular, 57 percent thought of horoscopes as "not at all scientific," while only 24 percent thought the same about astrology. At the same time, whereas 7 percent considered horoscopes as "very scientific," 26 percent expressed the same view for astrology. Interestingly, even though medicine, physics, and biology were considered the most "scientific" subjects, more people considered astrology scientific than considered history and economics scientific. History and economics were followed by homeopathy and horoscopes as the least scientific subjects. The rankings of these subjects as scientific were not the same among the various European countries. Overall, people in Eastern European countries were found to accept astrology more than those in Western European countries. The researchers also found that people often confused astrology with astronomy, as well as that the better one's understanding of scientific concepts and factual knowledge, the better one was able to distinguish science from pseudoscience.

Overall, it seems that at least in North America and Europe, people generally think of astrology as being a subject that is quite scientific. Worse than that, several people seem to believe that they can draw on astrology in order to make decisions about their lives. This is likely due to an underlying belief that there is a "scientific" way to look into one's destiny, and take the best actions. Several other studies have shown that people tend to intuitively think that life is driven by destiny. Thus, rare but important life events, such as finding the love of one's life or experiencing a tragic accident, are often perceived as predetermined, as "meant to happen." In many cases, such events are attributed to a god. For instance, from a survey of over one thousand US adults it was concluded that 56 percent of them completely or mostly agreed that "God is in control of everything that happens in the world"; 38 percent completely or mostly agreed that "earthquakes, floods and other natural disasters are a sign from God"; and 40 percent agreed that "natural disasters are God's way of testing our faith."[11] These people expressed the belief that significant life events, such as natural disasters, are not coincidental or simply natural but rather have a deeper underlying cause and purpose, being either signs from a god or a test of one's faith.

In a set of studies in Canada, researchers conceptualized fate attributions independently of particular cultures or religions, and also tried to disentangle them from other related ideas.[12] The aim of these studies was to investigate whether belief in God could explain differences in fate attributions among Christians and nonreligious people, as well as whether perceptions of causal complexity, the idea that multiple causes produce the same outcome independently or interactively, could explain differences between different ethnic/cultural groups. The 213 participants, who were either of European or Chinese ancestry and either religious or nonreligious, were given several scenarios, including the following—what would your answers be? Just circle those that you prefer.

1. It was 8:00 a.m. in the morning and the street was busy as usual. Kelly, on her way to school, stopped and reached down for her shoelace. While bent over she found a little diamond ring lying right in front of her which couldn't have been spotted otherwise.
   (a) It was a fluke that Kelly found the ring.
   (b) Kelly was meant to find the ring on that day.
2. Each day Jill activates the security alarm before leaving the house for work. Yesterday Jill was on the phone with a very important client, and she was so focused on the conversation that she totally forgot about putting the alarm on. Jill found out her house had been broken into when she came home. Her friends were surprised by the mishap because Jill is well-known for being a mindful person.
   (a) It was pure chance that Jill's house was robbed on that day.
   (b) It was the day her house was meant to be robbed.

3. Bob ran across his friend Jim and Jim's sister on the street one day. Later Bob and Jim's sister fell in love with each other and got married. Do you think:
   (a) It was pure chance that Bob ran across his friend Jim and his sister that day.
   (b) Bob and Jim's sister were meant to meet each other on that day.

4. John was walking in downtown when he spotted a Toonie lying on the ground. Just when John stopped and picked up the coin a window from an apartment above falls from its frame, hurting John severely.
   (a) John's injury could have been avoided if he happened to pass the coin instead of picking it up.
   (b) Even if John had not picked up the coin, he would have been injured by something else.

So, what did you choose? Would you agree with the idea that highly unlikely events such as those described above cannot happen just by chance? Would you rather agree with the idea that some intentional agent must be behind these events? Based on participants' responses, it was concluded that Christians made more fate attributions than the nonreligious, and that East Asian Canadians made more fate attributions than European Canadians. Therefore, East Asian Christians were the group showing the strongest tendency to attribute events to fate. This finding indicates that fate attributions do not follow a simple pattern along cultures or religions, as the combination of a Western religion (Christianity) and of a non-Western ethnic background (East Asian) seemed to produce the strongest tendency for fate attributions. Therefore, it's not as clear-cut as thinking that having a religion makes one more prone to fate attributions.

It might indeed be reasonable to think that religious belief in a benevolent creator makes people perceive plan and purpose in life events. However, things might be working the other way around. It could be the case that belief in destiny also emerges from teleological intuitions, which were already discussed in chapter 1. Convenient for our purposes here, there is a study that investigated exactly that. In particular, the researchers studied the teleological beliefs of God-believers and nonbelievers separately, in order to determine if there were any differences between the two groups. They also investigated whether teleological beliefs were stronger among God-believers than nonbelievers, in order to see whether religious belief might reinforce the idea that significant life events happen for a reason. Their first study involved one hundred American adults who were given the following definition of fate: "Many people believe that significant life events are meant to be and that they happen for a reason. They believe that there is an underlying order to life that determines how events turn out. These ideas are usually referred to as a belief in fate." Participants were then asked to state how strongly they believed in fate and whether they believed that fate is generally fair, kind, and instructive. As a group, most participants expressed some degree of belief in fate, with only about one out of five explicitly denying it. However, there were differences between believers and nonbelievers. Among those believing in God, the vast majority expressed some degree of belief in fate, whereas only half of nonbelievers did the same and a little less than half of them rejected it. It was also found that about two out of three believers and one out of five nonbelievers agreed that they sometimes see signs in significant life events, and that "everything works out for the best in the end"; and four out of five believers, and half of nonbelievers, agreed that there is "order in the universe." The conclusion that can be made from these results is that it is possible for one to see life events as "meant to be" even in the absence of a belief in God, as well as that the latter enhances this tendency. This in turn implies that teleological reasoning about life events is not exclusively due to religious views, but rather reflects the intuition to perceive purpose in the world.[13] Similar results and conclusions were reported by another research group that concluded that in addition to belief in God, individual differences in the tendency to think in teleological terms may enhance belief in the deeper meaning of significant life events.[14]

Another study aimed at investigating whether people tend to explain naturally occurring events in terms of inherent purpose and design, in the same way that they explain the occurrence and characteristics of natural objects (see chapter 1).[15] Participants included equal numbers of American and British atheists and theists. Those who agreed with the statement "I believe in a soul, spiritual force, or god(s)" were considered as theists, whereas those who agreed with the statement "I do not believe in a soul, spiritual force, or god(s)" were considered as atheists. These participants were interviewed and were asked to describe two different memories, a "learning experience" (any time they felt like they had learned something really important or mature as a person) and a "low point" (any time they felt great sadness or experienced a loss that left a lasting impression). Then they were asked several questions, including ones about whether they thought they had any kind of control over these events, or whether they thought of them as unexpected. Their responses were coded as teleological if participants had indicated that they perceived a supernatural agent having a particular intention related to an event; in contrast, if participants had explicitly rejected the idea that there was an intention or a final cause for an event, their responses were considered as non-teleological. Overall, 74.5 percent of non-teleological responses came from atheists and 81.7 percent of the teleological responses came from theists. However, 50 percent of the atheists gave at least one teleological response, which suggests, according to the researchers, that people may intuitively view important life events in teleological terms, without this necessarily being due to religious belief.

To investigate how fundamental these intuitions are and whether they might have roots in childhood, researchers studied the choices that children made between teleological and non-teleological accounts of significant life events.[16] The study involved five- to ten-year-old children and adults, and the researchers aimed at investigating whether the adult perception of purpose in significant life events has early roots in childhood, by asking participants to choose between two types of explanations. The events and the alternative explanations shown to participants are presented below. Why don't you take the test to see which explanations make more sense to you? Circle the ones that do!

1. Maggie's brand new house burned down in a fire because:
    (a) Maggie was playing with matches indoors.
    (b) it was meant to teach Maggie that it is dangerous to play with fire.

2. Luke got very sick and had to spend a whole month in the hospital:
    (a) because Luke caught a virus when he was traveling in another country.
    (b) in order to show Luke that people should be thankful when they are healthy.

3. Linda was teased on the playground by the other kids in her class:
    (a) because Linda's classmates are bullies.
    (b) so that Linda could learn how to be strong when dealing with mean people.

4. Monica lost her favorite stuffed teddy bear:
    (a) because Monica accidentally left her teddy bear on the school bus.
    (b) in order to make Monica learn to be more careful with her things.

5. Brianna's pet puppy ran away from home because:
    (a) Brianna left the front door open.
    (b) it was meant to teach Brianna that looking after a pet is a big responsibility.

6. Ben came in first place in the neighborhood track race because:
    (a) Ben practiced running more than any of his neighbors.
    (b) it was meant to show Ben that anything is possible if you work hard at it.

7. Sandy made a new best friend on her first day of school:
    (a) because Sandy and her new friend like all the same things.
    (b) in order to show Sandy that school can be a fun place

8. Ron got to see his favorite movie after waiting in line for hours at the theater because:

    (a) there were still movie tickets for sale when Ron got to the ticket stand.

    (b) it was meant to be a lesson that being patient is important.

9. Arnold was picked to be the star of a new children's television show because:

    (a) Arnold was a very good actor.

    (b) it was a sign that Arnold was meant to be famous.

10. Billy scored a goal on his very first day of soccer practice:

    (a) because Billy kicked the soccer ball right past the goalie's hands.

    (b) in order to show Billy that he should be a soccer player when he grows up.

So, what did you choose? The more (b) explanations you have chosen, the more you tend to think about purpose in life events. The study concluded that the younger children agreed with seven teleological explanations on average; the older children, with five teleological explanations on average; and the adults, with two teleological explanations on average. In addition, there was no effect of religious beliefs or practice on the test results. Therefore, one may conclude that there must be some connection between fundamental intuitions expressed from a very young age and the tendency to perceive purpose in life events. An independent study helped establish children's preference for teleological explanations.

The findings of this and the previous studies support the conclusion that it is not religious belief that makes one perceive purpose in life events. Whereas it is certainly the case that religious belief is compatible with, and perhaps enhances, the perception of purpose in life events, the latter is something that nonreligious people also exhibit. Therefore, it can be the case that it is the design stance that makes people perceive purpose in life events, and that religion simply adds to what is already there. A belief that is related to seeing purpose in life events, but which is also slightly different, is the belief in reincarnation, that is, the rebirth of the soul in a new body after death, or, more simply, the idea that we are born into this world again. The connection with destiny in this case is that one might think that people do not die, but it is their destiny to come back in this world in a different body. A study of US adults concluded that one out of five believed in reincarnation; one out of five were not sure what to believe about it; and three out of five stated that they did not believe in it. The number of Americans believing in reincarnation has been quite steady since 1990.[17] The situation seems to be quite similar in Europe. About one out of five people in Western European and Nordic countries seem to believe in reincarnation, with variations among the various countries. This belief was even higher in Eastern European countries, particularly in the Baltic countries.[18] These results are especially interesting, as the main European religions reject the idea of reincarnation, in contrast to most Indian religions such as Buddhism and Hinduism.

Even more widely accepted is the idea of life after death. In a survey conducted in several different countries, about half of the participants stated that they believe in some form of afterlife; for instance, about one out of five believed in an afterlife "but not specifically in a heaven or hell," and a similar number agreed that "you go to heaven or hell." In contrast, about one out of four stated that "you simply cease to exist," and a similar number that they "don't know what happens." The belief in an afterlife, but not specifically in heaven or hell, varied among countries and was highest in Mexico and Russia. The belief in going on to heaven or hell after death was more common among people from Indonesia, South Africa, and Turkey. Most people who answered that "you simply cease to exist" were from South Korea and Spain. It must be noted that fewer people than what was reported above were found to believe in reincarnation. Finally, most people who answered that they "don't know what happens" were from Sweden, Germany, and Japan.[19]

The idea of being reborn in this world or continuing one's life after death are both very tempting to believe. It is indeed difficult for many people to accept that one ceases to exist when the body dies, as this could be perceived as making life meaningless and allows for a nihilistic view of life. It is perhaps more comforting to think that our destiny extends beyond our current life to a life after death, or even better to new lives through reincarnation. Rather than thinking that life has no inherent purpose and that we are nothing but the outcomes of some cosmic coincidences, it is more intuitive—in accordance with the design stance—to think that someone (God, Nature, the Universe, or whatever you like) has a plan for us, which has no limits, and of which our current life is just a stage. This is why the continuation of life in some way—be it life after death or reincarnation—is an idea accepted and promoted by many religions across the globe.[20] If we believe in destiny, it makes more sense to believe that life does not end once we are biologically dead. Apparently, this idea assumes some kind of body-soul dualism that is subject to certain kinds of criticism.[21]

The design stance is quite evident in the explanations for life events that people gave in the research presented in this chapter. If something significant happens, whether bad or good, it is intuitive for many people to accept that there must have been a plan behind its occurrence. This is more likely found with religious people, but even those who do not believe in a supernatural deity tend to think like this, albeit to a lesser degree. It seems that many people cannot accept that significant events, such as finding the love of one's life or experiencing the death of the beloved person in a rare accident, simply happen. The problem is that we are trying to make sense of events that are senseless. Therefore, we look for a purpose behind them; this is the only way that we can make sense of them. Consider a plane crash. Millions of people fly in planes every year. So, when a group of people die in a plane crash, I cannot imagine any special reason that this happened to these specific people and not to others. This does not mean that there was no cause for the accident; there is always one or more causes. A certain airplane, but not others, crashed because of a pilot's mistake, an insufficient control on ground, or very bad weather conditions, for instance. Any of these could be *causes* of the accident. But this we know already. Nevertheless, we go beyond causation and ask questions such as why that particular pilot made the mistake, why the particular control was insufficient, why the particular weather conditions were that bad, and overall why those people had to die. Unfortunately, this is life; bad things happen. But more often than not, this seems like an insufficient answer to people who are trying to make sense of why something bad happened. Hence the common teleological belief that there must have been a purpose or plan in whatever happened.

In part 3 of the present book I explain that all life events can be causally explained after the fact, but whether events will take one or the other direction depends on local conditions, circumstances, and critical events, rather than on any underlying purpose or design. Things just happen, and I am using the story of the development of Charles Darwin's theory of evolution to illustrate this. This example has nothing to do with life or death, but in my view the principle I am describing applies in the same way to all life events. I chose this particular topic because it is related to biological science, which is the focus of the present book, because Charles Darwin wrote down his thoughts about his life events in his autobiography, because it is a story well studied by historians, and—perhaps mostly—because it is a topic easier to follow, as it does not concern the big questions of life that puzzle us. Having chosen this topic, I do not imply that people may have thought that Darwin was destined to do whatever he did. He is just an example that nicely illustrated the contingencies of life.

The important point here is that *On the Origin of Species*, presenting Darwin's theory of evolution, was not waiting out there to be written. Rather, Darwin's theory took the form that it had and it was published when it did because of particular turning points. These were identified not by me but by Charles Darwin himself (at least in the way I have interpreted what he wrote) in his autobiography. To show this, and to let Darwin speak for himself, and about the significance of those events, I have included rather extended quotations in part 3. As I show, Darwin could have published his theory at a different time, before or after 1859, and in a different format (e.g., a paper or a multivolume treatise). It seems that the format of *On the Origin of Species* was the most appropriate one, but as I show this was due to particular turning points. I believe that this well-studied case, which is free from any metaphysical thoughts and implications, shows how life events and personal decisions in particular contexts affect life outcomes, in this case the development and the publication of a groundbreaking theory. The form in which this theory was published had nothing to do with any kind of destiny.

Chapter 4

"GOD'S WISDOM REVEALED IN HIS CREATION"

Before the publication of Charles Darwin's *On the Origin of Species by Means of Natural Selection* in 1859, many people accepted the special creation of species as a fact, that is, they accepted that species were created in their present form, although the idea of evolution was already in the air. Two of Darwin's achievements were to provide convincing arguments and evidence that evolution occurs, as well as a natural process— natural selection—as a means of how it might occur. His theory was not immediately accepted, but the idea of evolution became more and more prevalent toward the end of the nineteenth century. Developments in genetics in the beginning of the twentieth century came to complement Darwin's initial theory, eventually producing what is generally known as the neo-Darwinian theory of evolution. Today this theory is considered the central unifying theory of biology, according to the vast majority of practicing scientists. But at the same time its acceptance by the public is not equally high. A main reason for this is that it is perceived to be in conflict with religious worldviews that consider a supreme being as the creator of life, especially of humans.

People objecting to the theory of evolution and accepting the idea of the special creation of organisms, particularly humans, are usually described as creationists. In particular, creationism is the belief that a god created the universe through a series of miraculous interventions. Young Earth creationists believe that the world was created in six days of twenty-four hours each, at some point during the last ten thousand years, whereas old Earth creationists accept the scientific account of the age of the Earth but still believe in the miraculous creation of life and organisms. The more modern, and seemingly more sophisticated, approach to the study of creation is generally described as "intelligent design theory," even though variations exist. Overall, the central point made by intelligent design (ID) proponents is that natural processes are not the most plausible explanations for the complexity we see in organisms. They criticize evolutionary theory, arguing that it cannot explain the characteristics of organisms that we observe. One of the arguments is that complex organs, such as the vertebrate eye or the bacterial flagellum, are irreducibly complex systems—systems that become nonfunctional if one of their parts is removed. Therefore, such systems cannot have gradually evolved through natural selection and can only have been created by a divine designer. Hereafter, I refer collectively to these views, creationism and intelligent design, as IDC (intelligent design creationism), for the sake of simplicity.[1]

There exist several institutions that promote this kind of thinking, especially in the United States. One of these is the Institute for Creation Research, which "has equipped believers with evidence of the Bible's accuracy and authority through scientific research, educational programs, and media presentations, all conducted within a thoroughly biblical framework."[2] This is one among many different initiatives to advance the view that the world and humans have come into being through the actions of the Judeo-Christian God, and that whenever we study nature all we find is evidence for these actions: "In particular, the cause of our universe coming into being, and of its continuing to operate as it does, is a dynamic display of the Creator's wisdom, some of which we can scientifically discover and understand. When we do, it is like walking in the footprints of someone who previously walked through a snowdrift."[3] According to IDC proponents, nature provides evidence for the existence of God, as only he can be the ultimate cause of the magnificent complexity we see in the living world. It is improbable, if not impossible, the story goes, for all of the complex features that we find in organisms to have simply evolved through natural processes and without any designing mind behind them. A popular response to this view has been that there indeed exists a designer, albeit not the kind of designer IDC believers expect: "Natural selection, the blind, unconscious, automatic process which Darwin discovered, and which we now know is the explanation for the existence and apparently purposeful form of all life, has no purpose in mind. It has no mind and no mind's eye. It does not plan for the future. It has no vision, no foresight, no sight at all. If it can be said to play the role of watchmaker in nature, it is the blind watchmaker."[4] This is an example of how scientific responses to IDC views aim at addressing their religious aspect but not the design one. Evolution, particularly human evolution, is perhaps the case where the design stance is most prominent. But let us first see certain IDC arguments in some detail.

A well-known book that aims at providing scientific evidence for the argument from design, that is, that life exhibits design and therefore there must exist a designer, is Michael Behe's *Darwin's Black Box*.[5] In this book, the author proposed the concept of irreducible complexity that, he argued, makes gradual evolution by natural selection conceptually impossible. According to Behe, an irreducibly complex system is

> a single system composed of several well-matched, interacting parts that contribute to the basic function, wherein the removal of any one of the parts causes the system to effectively cease functioning. An irreducibly complex system cannot be produced directly (that is, by continuously improving the initial function, which continues to work by the same mechanism) by slight, successive modifications of a precursor system, because any precursor to an irreducibly complex system that is missing a part is by definition nonfunctional.[6]

To put it simply: an irreducibly complex system is a functional system that becomes nonfunctional if we remove one of its parts. And, the argument goes, if this is the case, then such a system cannot have been produced by the gradual modification of earlier systems because these would have to be, by (Behe's) definition, nonfunctional as well and thus could not have existed. This, according to Behe, would pose problems to accepting the theory of evolution:

> An irreducibly complex biological system, if there is such a thing, would be a powerful challenge to Darwinian evolution. Since natural selection can only choose systems that are already working, then if a biological system cannot be produced gradually it would have to arise as an integrated unit, in one fell swoop, for natural selection to have anything to act on.[7]

To make his point, Behe quoted Darwin's *On the Origin of Species*: "If it could be demonstrated that any complex organ existed, which could not possibly have been formed by numerous, successive, slight modifications, my theory would absolutely break down."[8]

We can summarize Behe's argument: An irreducibly complex system is a functional system consisting of components A-B-C-D that becomes nonfunctional if any of the components A or B or C or D is removed. Evolution by natural selection is based on the assumption that system A-B-C-D must have evolved from a simpler system such as A-B-C or B-C-D or A-C-D or A-B-D, and so on. But any system such as A-B-C or B-C-D or A-C-D or A-B-D, and so on should by definition be nonfunctional; and nonfunctional systems are not selected in the course of evolution. Therefore, the evolution of an irreducibly complex system by means of natural selection is conceptually impossible.

The question then becomes how one can identify an irreducibly complex system. According to Behe, the first step should be to specify both the function of the system and all its components, under the assumption that all the parts of such a system contribute to this function. In order to make his point, Behe refrained from using a complex biological example and chose a simple mechanical one: the mousetrap.

> The function of a mousetrap is to immobilize a mouse so that it can't perform such unfriendly acts as chewing through sacks of flour or

electrical cords, or leaving little reminders of its presence in unswept corners. The mousetraps that my family uses consist of a number of parts . . . : (1) a flat wooden platform to act as a base; (2) a metal hammer, which does the actual job of crushing the little mouse; (3) a spring with extended ends to press against the platform and the hammer when the trap is charged; (4) a sensitive catch that releases when slight pressure is applied, and (5) a metal bar that connects to the catch and holds the hammer back when the trap is charged. (There are also assorted staples to hold the system together.)[9]

The second step in identifying an irreducibly complex system would be to examine if all the components are required for the function. In the case of the mousetrap, the answer is clearly yes, according to Behe, because if the wooden base was missing, there would be no platform for the attachment of the other components; if the hammer was missing, the mouse would not be trapped; if the spring was missing, the hammer and the platform would jangle loosely and again the mouse would not be trapped; if the catch or the metal holding bar was missing, then the hammer would not stay open because the spring would snap it as soon as one let it go. Therefore, according to Behe, the mousetrap is an irreducibly complex system and it cannot have evolved in gradual steps by some precursor missing either the wooden platform, or the metal hammer, or the spring, or the catch of the metal bar, because such a system would be nonfunctional and thus useless.[10] Behe also extended his argument to biological structures such as the bacterial flagellum, in order to show that such systems are also irreducibly complex and so cannot have evolved by natural selection. His conclusion about the failure of the idea of evolution by natural selection seems definitive:

> In summary, as biochemists have begun to examine apparently simple structures like cilia and flagella, they have discovered staggering complexity, with dozens or even hundreds of precisely tailored parts. It is very likely that many of the parts we have not considered here are

required for any cilium to function in a cell. As the number of required parts increases, the difficulty of gradually putting the system together skyrockets, and the likelihood of indirect scenarios plummets. Darwin looks more and more forlorn. New research on the roles of the auxiliary proteins cannot simplify the irreducibly complex system. The intransigence of the problem cannot be alleviated; it will only get worse. Darwinian theory has given no explanation for the cilium or flagellum. The overwhelming complexity of the swimming systems push us to think it may never give an explanation.[11]

The power of Behe's argument was admired by other major figures in the IDC camp, such as William Dembski, who supported it and tried to further defend it, emphasizing the inadequacy of Darwinian evolution:

In order even to use the origination inequality, one must first propose specific evolutionary pathways leading to irreducibly complex biochemical systems like the bacterial flagellum. Only with such proposals in hand can one begin to estimate the probabilities that appear in the origination inequality. Moreover, once such proposals are made, they invariably point up the inadequacy of the Darwinian mechanism because the origination probabilities associated with irreducibly complex biochemical systems have, to date, always proven to be small. Design theorists take this as strong confirmation that these systems exhibit specified complexity and are in fact designed. Darwinists, by contrast, take this as simply showing that evolutionary biology has yet to come up with the right evolutionary pathways by which the Darwinian mechanism produced the systems in question.[12]

Later in his book Behe expressed his admiration for the theologian William Paley, and for his attempt to present the argument from design, which he considered the best pre-Darwinian one. Paley started his book *Natural Theology* by comparing a stone and a watch. His argument was that if he came across a stone, he would not wonder how the stone came to be there but would rather think that it had been there forever. However, he would not think the same for a watch, had he come across one. The reason for this is that the parts of a watch are formed and adjusted in a particular way in order to indicate what time it is. If the parts of the watch had a different shape or size, or if they were placed in another manner or order, the watch would not be functional. Therefore, a watch exhibits intelligent and intentional design, which in turn implies the existence of a watchmaker who designed the watch in a particular way so that it would be functional. The watchmaker had the intention to make an artifact that tells the time, and he implemented a particular design that involved a specific arrangement of its parts, which in turn stands as evidence for his/her existence. In contrast, the stone has no such features, and no inference to a stone maker makes any sense.[13] There are important differences between Paley's views and those of IDC proponents, but it is interesting that the latter view Paley as their intellectual forefather.[14] Differences notwithstanding, IDC proponents see Paley as providing evidence from nature for the existence of God, and they believe they are doing the same.

IDC is quite prevalent in the United States, where for at least the past thirty-five years over 40 percent of people have agreed with the idea that God created humans in their present form. During the same time, over 30 percent of people have accepted the idea of evolution—as long as it is guided by God. Therefore, less than 20 percent of Americans— but with an increasing tendency in more recent years—have accepted the idea that humans have evolved and that God had no part in this process. The most recent poll took place in May 2017 with a sample of over one thousand US adults. It was found that 38 percent accepted the idea that God created humans in their present form; 38 percent, that God guided human evolution; and 19 percent, that humans have evolved without any divine intervention. It must be noted that this is the first time in the history of this poll (since 1982) that belief in the direct creation of humans by God was not the most common response. The IDC view was accepted more so by the more religious and less educated people.[15] Compared to the 2014 poll, the amount of people accepting evolution without God was exactly the same; also, more people accepted evolution guided by God, and less accepted the idea that God had created humans. An interesting finding, except for the higher acceptance of evolution in 2014 compared to the past, was that more Americans self-reported familiarity with evolution (79 percent) than with creationism (76 percent). Nevertheless, those familiar with evolution did not necessarily accept the evolutionary perspective that does not require the involvement of a god; 64 percent of those who were very familiar with the theory of evolution chose either this explanation or the one that humans evolved with a god guiding. The conclusion made by the researchers was that it was not clear that attaining more education would produce any change in viewpoint, as many Americans accept creationism mostly because of their religious beliefs.[16]

Of course, the United States is not the only country in the world in which IDC is quite commonly accepted. It is also prevalent in other countries, such as Russia, Brazil, Turkey, South Africa, Indonesia, and Saudi Arabia. A global survey conducted in twenty-three countries and involving nearly nineteen thousand adults concluded that 51 percent of global citizens definitely believe in a god or a supreme being, ranging from 4 percent in Japan to 93 percent in Indonesia. The survey also concluded that, globally, 28 percent believe in creationism, ranging from 8 percent in Belgium to 75 percent in Saudi Arabia; furthermore, it found that 31 percent are "unsure what to believe" about evolution, ranging from 18 percent in Saudi Arabia to 40 percent in Russia. Of course, as expected, the amount of people accepting IDC was higher in those countries in which more people described themselves as religious. But if you look at the numbers carefully, you can make an important primary conclusion that not all religious people accept IDC; on average 51 percent of global citizens believe in God, but only 28 percent of them are creationists. Most important, we should keep in mind that approximately one in three people in most of the twenty-three countries studied was unsure what to believe about evolution.[17]

Another important point to keep in mind is that survey questions about the public acceptance of evolution may have been formulated in ways that could lead to biased responses and distorted representations of the actual views held by participants. This can happen, for instance, if the questions focus on human evolution or get into too much detail about personal beliefs and worldviews, as doing so can potentially make respondents feel uncomfortable and thus not likely to express their actual views.[18] At the same time, polls that ask participants to choose among particular responses may force them to give responses they would not otherwise give. For instance, in one poll on evolution, participants were asked to choose among three statements: one about "evolution theory" suggesting that humans have developed over millions of years from less advanced forms of life, without any intervention from God; one about "creationism theory" suggesting that God created humans once in their present form at some point during the last 10,000 years; and the "intelligent design" theory suggesting that the features of organisms are best explained by the intervention of God. In this case, these three options might eventually force religious people to select either the creationism or the intelligent design response, as there was no clear option, for instance, about evolution being initiated or guided by God. As a result, the amount of people holding IDC views might be exaggerated, and so the results of such studies would be misleading.[19] What is to be gleaned from all of this? In sum, religiosity does not necessarily entail acceptance of IDC views, and the two should be clearly distinguished.

For these reasons, you should be very careful in interpreting polls on evolution, as it is possible that the way the questions are posed to participants may have an influence on the results. For instance, two polls conducted in the United States during 2014 by Pew Research Center and Gallup included questions about evolution framed in different terms, as shown in table 4.1.

Type of question	Answer	2014
Evolution through natural processes	"Human evolved but God had no part in process" (Gallup)	19%
	"Humans have evolved due to natural processes" (PEW)	35%
God involved in evolution	"Humans evolved with God guiding" (Gallup)	31%
	"Supreme being guided evolution" (PEW)	24%
Creation	"God created humans in present form" (Gallup)	42%
	"Humans have existed in present form since beginning" (PEW)	31%

Table 4.1. Different results in two studies conducted in the United States, asking essentially the same questions but framed in a different manner (data from the 2014 PEW and Gallup polls).

If you compare the responses to the two polls, you will see that almost twice as many participants selected evolution through natural processes in the Pew poll (35 percent) than in the Gallup poll (19 percent). Similarly, more participants selected an option explicitly or implicitly involving God, either guiding evolution or creating humans, in the Gallup poll (73 percent) than in the Pew poll (55 percent).[20] Clearly, there are some methodological issues to consider, but it seems likely that reading the word "God" in the items of the Gallup poll but not in the items of the Pew poll had an influence on participants' responses.

The important question for our discussion here is whether it is religious belief in a benevolent divine creator that makes IDC proponents accept the idea that organisms, and especially humans, were specially created and intelligently designed. A relation seems to exist. But does belief in a god make these people perceive the living world as designed? Or is it the perception of design in the living world that compels people make the inference to a god as the designer? These relations are not equivalent; in the former case, it is belief in a god that comes first and supports the conclusion that the world must be designed; in the second case, it is the perception of design in the living world that comes first and then the inference to a god as designer comes as a second step. In the first case, religious belief could be thought of as the cause for accepting IDC views, whereas in the second case religious belief is simply compatible with another view, the perception of design in the living world that exists independently. But let us now see more closely how religiosity and acceptance of IDC may be related.

A study of 185 children in the United States investigated whether children from Christian fundamentalist school communities expressed more creationist views on the issue of the origin of species than did children who came from nonfundamentalist school communities. One could reasonably expect more children from the former group to express creationist views compared to children from the latter group. Participants (aged five to thirteen) were divided into three groups based on their age. In all groups, more students from fundamentalist school communities provided creationist explanations to all tasks. In contrast, students from nonfundamentalist school communities provided explanations that were different in different age groups. However, students aged 8–10 from nonfundamentalist school communities also provided mostly creationist explanations. This is a very interesting finding that shows that it is not just the religious background of the family that may have an influence on children's beliefs (since children at this age from both backgrounds preferred creationist explanations for the origin of species). It should also be noted that all participants had mixed views.[21] The result that the preference for creationist explanations was not restricted to children from fundamentalist religious backgrounds but was also found in children from nonfundamentalist backgrounds supports the conclusion that this preference is not due to religion (at least not only) but perhaps due to a fundamental teleological intuition, that is, an intuition about the existence of purpose and design in nature.

Another study aimed at further investigating this issue. Based on the assumption that the United States and Great Britain share many cultural characteristics but differ in religiosity, with the British being less religious than Americans, the study involved forty-eight American and forty-eight British children, all between the ages of seven and ten. The main aim of the study was to investigate whether there was a difference in the preferences of American and British children for teleological explanations. Although there were certain differences between the two groups, overall the explanations of students of both groups were quite similar. They generally preferred teleological explanations both for nonhuman organisms and for nonliving natural objects. This finding is interesting if we accept that British children are less likely to be exposed to cultural influences about intention or design in nature than are American children. The results of this study suggest that the preference for intentional explanations (creationist or teleological-functional) may not just be the outcome of cultural-religious influences but may in fact be due to a more fundamental, human intuition. Because of that, it seems that children may require only a minor external influence to endorse teleological explanations for natural phenomena.[22]

This being stated, it is not necessarily the case that a teleological explanation requires a creationist view. Whereas one can provide teleological explanations for natural phenomena as a result of one's acceptance that these phenomena were created by a god, it is also possible to provide teleological explanations because one perceives natural purposes, that is, goals and functions that serve some role, and then make the leap to thinking that these might be the outcome of conscious design. A subsequent study aimed at exploring this issue further, by investigating whether children's tendency to explain natural phenomena in terms of purpose and their beliefs about the existence of intelligent design in nature were related. Fifty-five British elementary school children participated in this study. Among others, they were given the following tasks. Why don't you take a look at them, and note which responses you would endorse?

1. Why did the first ever thunderstorm occur?
    (a) The first ever thunderstorm occurred because cold and warm air all rubbed together in the clouds.

    (b) The first ever thunderstorm occurred to give the earth water so everything would grow.

2. Why did the first ever flood occur?
    (a) The first ever flood occurred because it rained so much that water covered everything.
    (b) The first ever flood occurred to wash and clean the earth of bad things.

3. Nonliving natural objects
    Why did the first ever river exist?
    (a) The first ever river existed because big blocks of ice melted and made lots of water.
    (b) The first ever river existed to provide fish and crocodiles with somewhere to live.

4. Why did the first ever mountain exist?
    (a) The first ever mountain existed because a volcano erupted and cooled into a big lump.
    (b) The first ever mountain existed to give animals a home and somewhere to go climbing.

5. Why did the first ever monkey exist?
    (a) The first ever monkey existed because some animals developed into people and some developed more like this.
    (b) The first ever monkey existed to give trees a swinging animal and tigers something to eat.

6. Why did the first ever bird exist?
    (a) The first ever bird existed because an animal that lived on the ground began to develop wings and fly.
    (b) The first ever bird existed to eat worms and insects so there wouldn't be too many of them.

7. Why did the first ever boat exist?
    (a) The first ever boat existed because bits of wood got fixed together.
    (b) The first ever boat existed to carry people over water.

8. Why did the first ever hat exist?
    (a) The first ever hat existed because material got formed into a round shape.
    (b) The first ever hat existed to keep someone's head warm.

If you selected more (b) than (a) answers, then you are also inclined toward teleological explanations. However, these are legitimate only for questions 7 and 8, which refer to artifacts. After answering the above tasks, children were asked explicitly, for each one of these objects in the same order, if someone or something made these or if they just happened to exist or occur. Children could thus state that the objet under question was either made by an intentional agent ("someone") or a nonintentional agent ("something"), or whether it just spontaneously emerged. The researchers concluded that there seemed to exist a systematic connection between teleological explanations and intuitions about intelligent design in nature.[23] This supports the idea that it could be teleological intuitions that produce IDC views, and not religious belief—or at least not religious belief alone. This entails once again that belief that purpose and design may be something fundamental in human cognition, and that IDC views stem from that. Religion follows, entering the scene during one's life, and is of course compatible with these views. In other words, creationists may at first be intuitive teleologists who then become religious. This entails that even nonreligious people might perceive purpose and design in nature.

Another study provided additional evidence for exactly that. Its aim was to investigate whether nonreligious adults, who explicitly rejected religious belief, nevertheless exhibited the tendency to perceive nature as the intentional product of some agent.[24] The first part of this three-part study involved 352 North American adults, 225 of whom were categorized as "religious" and 127 of whom were categorized as "nonreligious" on the basis of their own rating across a scale of religious belief. They were randomly assigned either to a speeded or an un-speeded condition; then they were shown 120 pictures, and were asked to state whether or not "any being purposefully made the thing in the picture." From the pictures shown, forty were about phenomena related to organisms and nonliving objects (e.g., giraffe, maple tree, tiger's paw, mountain, stalagmite, hurricane) and the remaining eighty were included to track participants' understanding of the instructions, their abilities to respond at speed, and their possible response biases. The results showed that under speeded conditions even nonreligious participants had the tendency to think of natural entities as purposefully created; working under the time constraint does not allow time for reflection, therefore, these findings further support the conclusion that perceiving purpose in nature stems from cognition and not from religious belief. The researchers repeated the study a second time, with a group of 148 North American atheists. The results obtained were quite similar, as these atheists also had the tendency to perceive natural entities as purposefully created by some non-human being when they had to answer under speeded conditions. Finally, the researchers conducted a third study with 151 explicitly nonreligious Finnish adults in order to see whether the same results would be obtained with people outside America. It was indeed found that these explicitly nonreligious Finnish people also tended to view both living and non-living natural phenomena as purposefully made by a nonhuman being when they were tested under speeded conditions. Overall, the findings of these three studies supported the conclusion that humans may have a deeply rooted tendency to perceive nature as designed.

All of these studies suggest that it may not be religion but rather their intuitive preference for teleological explanations that makes people perceive organisms as created. It is these intuitions that in turn may make them think of organisms and nature more broadly as an artifact of divine design. This means that whereas one might plausibly think that people become religious and then see nature as the artifact of a divine creator, it could be the case that people tend to think of nature as an artifact anyway, and then religious belief simply fits well because it is perfectly compatible with our intuitions. In other words, our intuitions about purpose and design make us perceive organisms as having complex structures. They thus remind us of artifacts with which we are familiar (such as mouse-traps and watches), and so we apply artifact-based thinking to them. This thinking produces the need to look for a very competent designer, at least one more competent than humans, and it is in this way that the inference to a god is finally made. We intuitively perceive design in nature, and IDC views therefore seem to be intuitive. This tendency seems to diminish as people grow up, without ever being completely overwritten, with religious people maintaining it more strongly than nonreligious people.

Nevertheless, and no matter how intuitive they may seem to be, the arguments of IDC proponents have been criticized in various works by various scholars. A prominent response, which has been criticized by Behe himself and other IDC proponents, was given by biologist Kenneth Miller, who relied on an argument and a series of designs by biologist John McDonald. McDonald initially showed how one could start from Behe's five-part mousetrap and by removing the parts one after the other arrive to a one-part mousetrap: the catch, the hold-down bar, the hammer, and finally either the string or the platform. Each time the mouse-trap would be less and less effective, but it would still work.[25] After Behe's initial reaction, McDonald further elaborated his argument, also reversing it to show how a five-part mousetrap could be gradually built by a one-part one.[26] Kenneth Miller took the argument even further by showing that the one-part, two-part, three-part, and four-part objects could be functional, not necessarily as mousetraps but for different purposes. For example, the three-part object consisting of the platform, spring, and hammer could be a spitball launcher or a tie clip. If then one detached the spring from the hammer, the latter could be used as a keychain and the two-part object (platform and spring) could be used as a paper clip. Finally, the catch alone could be used as a toothpick, or the base as a paperweight. Miller provided a similar explanation for the bacterial flagellum.[27] His main point was that during evolution the function of the various intermediate forms might be different than that of the complete form, but they could nevertheless be favored by natural selection.

These arguments allow us to significantly revise Behe's argument, and actually turn it to a pro-evolution one. An irreducibly complex system is a functional system consisting of components A-B-C-D that becomes nonfunctional if any of the components A or B or C or D is removed. Evolution by natural selection is based on the assumption that system A-B-C-D must have evolved from a simpler system such as A-B-C or B-C-D or A-C-D or A-B-D, and so on. Any system such as A-B-C or B-C-D or A-C-D or A-B-D, and so on can have a function that might be different from the function of A-B-C-D; and functional systems are selected in the course of evolution. Therefore, a complex system does not have to be an irreducibly complex one, and the evolution of a complex system by means of natural selection is conceptually possible.

The above shows that there is no need to rely on design in order to explain the properties of complex systems. The analysis also shows how problematic machine metaphors can be when discussing evolution. Although they can be quite useful for explaining how biological systems work, we should always keep in mind that the use is metaphorical. Therefore, we should think of, for instance, the bacterial flagellum *as if it were* a machine, and not *as if it is* a machine. As already explained in chapter 1, artifacts such as mousetraps and airplanes are intentionally and intelligently designed for a purpose. Their intended use is what determined their properties, which emerge from the functional arrangement of their parts that is intended to achieve that use. But this is not the case for organisms, which are the product of natural processes and not an intelligent designer. It is exactly because organisms are not intentionally and intelligently designed that they have various traits that are useless or even disadvantageous, and which are better explained as outcomes of evolution rather than of design.

The results from polls presented earlier in this chapter showed that many people all over the world think that a god either created humans in their present form or guided our evolution. It likely seems inconceivable to them that our species may have evolved through natural processes. Indeed, we have so many distinctive features, such as standing on our hind limbs, having large brains for our body size compared to other animals, and—perhaps most distinctively—having a culture that includes numerous skills, such as having a language and thus being able to read and write a book like this one. All of these distinctive features may seem to be too special to have evolved by natural processes. This is why in part 4 I focus on human evolution to present some of its turning points, which brought us humans to our current condition. There I explain that, no matter how special our features seem to be, they can have plausibly evolved naturally, and therefore there is no need to attribute them to any divine designer. I also show that these features occasionally have bad consequences, which would be embarrassing for any intelligent designer we might like to credit for creating humans. The impact of contingencies on human evolution is rather difficult to conceive and realize. But the currently available evidence supports the conclusion that it was several turning points, and not intelligent design, that guided human evolution.

The research presented in this and the previous chapters supports the conclusion that we are inclined to intuitively explain life events, forms, and outcomes in terms of goals, purposes, and intentions. It should be by now clear that we find the idea of design very intuitive and that people can easily adopt it to explain outcomes in human development (chapter 2), life (chapter 3), and evolution (the present chapter). This tendency, which I describe as the design stance, prevents us from realizing the important role of critical events in shaping outcomes. Just as we intuitively tend to perceive design in the world, in hindsight we interpret the complexity of developmental, life, and evolutionary outcomes as if it were designed. Our intuitions thus blind us to the alternative paths that development, life, and evolution could have taken. In the rest of this book I use particular, representative examples to show the importance of turning points in these historical processes. Their impact is often hard to realize, and quite often it is underestimated. My intention is to bring those to your awareness and question the interpretations of human development, life, and evolution outcomes in terms of genetic fatalism, destiny, and intelligent design, respectively. Such interpretations cannot hold, I argue, if one appreciates the impact of turning points in human development, life, and evolution.

Part 2

TURNING POINTS IN
HUMAN DEVELOPMENT

Chapter 5

YOUR 70 TRILLION POSSIBLE SIBLINGS

Have you ever wondered if you were meant to come to this world? If you were born in order to achieve whatever you have achieved? Or are you here just by accident? By pure luck? Was there a plan for every one of us, or were we the outcome of particular circumstances? I remember my thoughts before each of my children were born—"Will it be a boy or a girl?" "What will it look like?" Of course, I had no way to answer many of these questions before the children were born. Now that both of them are over ten years old, I have answers to these questions. More than that, they seem as if they were meant to be around and enrich the lives of their family and friends in the way they do. This is my conclusion because I think of our common life and my experiences with them. I cannot imagine having children who would look different; or who would behave differently. Nevertheless, my aim in this chapter is to explain why my children were not meant to be born; neither was I, nor were you. I could not have predicted whether each of my children would be a boy or a girl, or what they would look like. Neither could my parents have done the same for me; nor your parents for you. Let us see why the fact that all of us are here is the outcome of particular turning points: critical events that are contingent *per se* and *upon* which future events are also contingent.

Whether a child will be a boy or a girl and how it will physically look depends largely on the combination of genetic material that it will receive from its parents. This is why siblings differ from each other, sometimes slightly, other times significantly. Of course, when we make such comparisons we look at only a handful of biological features such facial characteristics or height. But there are so many others. Nevertheless, most of what we look like—from basic, robust features (e.g., two eyes, two legs, two arms, two ears, etc.) to more variable ones (e.g., eye, hair and skin colors, shape of eyes and ears, height, etc.)—depends largely on how the information encoded in our genome is read, interpreted, and used within our cells, under particular intracellular, intercellular, and external environmental conditions during our development. The external conditions can be highly variable but are always within a certain range that allows life to exist. However, the variability in the combinations of the genetic material of two parents is enormous and it might produce very different siblings. As I explain in the next two chapters, how we look or whether we develop a disease depends on a lot more than our genome. But leaving these other factors aside, in this chapter I want to highlight the wide range of possible genetic combinations that are available before our conception.

Let us start with the fundamentals. In the nucleus of each cell of every one of us there normally exist 46 molecules of DNA (deoxyribonucleic acid). As you probably know, this is the molecule in which genetic information is encoded. This means that DNA is an information resource for the cell, on which the latter draws during development. DNA is not lying naked in the nucleus but is packed with proteins called histones. It thus forms a structure called chromatin, which is a highly condensed structure, resulting in the relatively long DNA molecules being packed within the tiny nucleus of each cell. During cell division, chromatin is condensed even more and forms 46 structures called chromosomes, which are clearly visible with a microscope. These 46 chromosomes are not entirely different from one another; for each one of them there is another one that is quite similar, and we thus have 23 pairs of similar chromosomes, which we usually describe as pairs of homologous chromosomes. These pairs have names that correspond to their characteristics when they are ordered from the longer to the shorter ones (forming a representation that we call a karyotype): "chromosome 1," "chromosome 2," and so on until "chromosome 22." In each of these 22 pairs of chromosomes, called autosomes, both chromosomes are approximately the same size. The last pair includes chromosomes that differ in size and that relate to sex development, which are called "chromosome X" and "chromosome Y"—we usually refer to these collectively as the sex chromosomes. Females have two X chromosomes, whereas males have an X and a Y chromosome.

It is important to note that chromatin and chromosomes do not only differ in how condensed they are. Whereas they do not differ qualitatively, in terms of the DNA that they contain, chromatin and chromosomes differ quantitatively, as the latter consists of twice as many DNA molecules as the former. This occurs because of DNA replication that takes place before cell division. We have never seen how DNA looks, but James Watson, mentioned in chapter 2, and Francis Crick long ago developed a model for its structure: the double helix model of DNA. According to this model, a DNA molecule consists of two strands linked to each other. Each of these strands in turn consists of repeating molecules (called nucleotides) that differ only in which of four "bases" they contain: adenine (A), thymine (T), cytosine (C), and guanine (G). The nucleotides of the same strand are linked in a linear sequence, whereas the nucleotides of the opposite strands are also linked, but in a specific manner: an A with a T, and a C with a G. As a result, there exist only two possible pairs of nucleotides (A-T and G-C), and so the two strands are complementary, that is, if there is a certain base at a certain point of one strand, then the complementary base should be on the respective point on the other strand. For example, if the sequence in one strand is GATTACA, then it can be easily inferred that the sequence in the respective position in the other strand is CTAATGT (because A forms a pair with T, and G with C).

This property of DNA also forms the basis for its replication. Various enzymes "open" and "read" the double helix of DNA. The GATTACA strand forms the basis for the synthesis of a new CTAATGT one. Similarly, the CTAATGT strand forms the basis for the synthesis of a new GATTACA strand. Thus, from an initial DNA molecule two new ones are formed that should be identical to each other (see figure 5.1a). As a result of this, each chromatin filament turns to a chromosome consisting of two sister chromatids, each of which consists of two identical DNA molecules (figure 5.1b). I must note that the transformation of a chromatin filament to a chromosome with two sister chromatids is more accurately represented in figure 5.1c. However, for the sake of simplicity, in this book I use the representation in figure 5.1b. I must also note that because they are qualitatively the same, we can refer to chromatin filaments (which is the form that exists for most of the life of a cell) as chromosomes as well.

Now, and this is the interesting part for our discussion here, each chromosome of each pair comes from each one of our parents. Therefore, the 46 chromosomes that I have in each one of my cells, are a paternal and a maternal one of each of the 23 kinds of chromosomes. In other words, for each of my 23 pairs of chromosomes, one chromosome was derived from my mother and the other was derived from my father. I therefore have a "chromosome 1" from my father and a "chromosome 1" from my mother, a "chromosome 2" from my father and a "chromosome 2" from my mother, and so on until "chromosome 22," whereas I have a "chromosome X" from my mother and a "chromosome Y" from my father. Similar is the case for my brother. However, why aren't we identical, as we have inherited chromosomes from the same parents? The answer is simple: we did not inherit the same chromosomes. Instead, we have each inherited a different combination of our parents' chromosomes, and not the same 23 chromosomes from each one of them. To understand how this is possible, our grandparents have to be considered, too. Let us call our paternal grandfather and grandmother PgF and PgM, respectively; and our maternal grandfather and grandmother, MgF and MgM respectively. Our mother thus has 23 MgF chromosomes and 23 MgM chromosomes, derived from each of her parents; similarly, our father has 23 PgF chromosomes and 23 PgM chromosomes, also derived from each of his parents. Let us for simplicity represent three pairs of chromosomes of our parents:

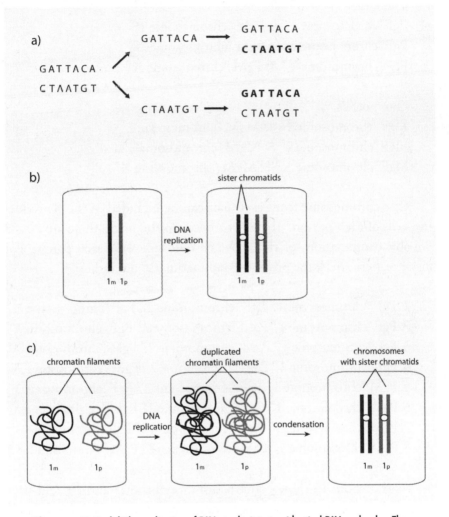

Figure 5.1. (a) The replication of DNA results into two identical DNA molecules. The newly synthesized strands are highlighted in bold. (b) As a result of this, each chromatin filament turns to a chromosome consisting of two sister chromatids, that each consists of a DNA molecule that is normally identical to that of the other. (c) This happens after both DNA replication and condensation of the chromatin to a more compact form.

Our father:
PgF "chromosome 1" & PgM "chromosome 1"
PgF "chromosome 2" & PgM "chromosome 2"
PgF "chromosome Y" & PgM "chromosome X"

Our mother:
MgF "chromosome 1" & MgM "chromosome 1"
MgF "chromosome 2" & MgM "chromosome 2"
MgF "chromosome X" & MgM "chromosome X"

One chromosome from each pair can be included in the reproductive cells of each person. Therefore, if you do the math, there are $2^3 = 8$ possible combinations of the above chromosomes for each parent. For instance, here are all the possible combinations in our father:

1. PgF "chromosome 1," PgF "chromosome 2," PgF "chromosome Y"
2. PgF "chromosome 1," PgM "chromosome 2," PgF "chromosome Y"
3. PgF "chromosome 1," PgM "chromosome 2," PgM "chromosome X"
4. PgF "chromosome 1," PgF "chromosome 2," PgM "chromosome X"
5. PgM "chromosome 1," PgF "chromosome 2," PgF "chromosome Y"
6. PgM "chromosome 1," PgF "chromosome 2," PgM "chromosome X"
7. PgM "chromosome 1," PgM "chromosome 2," PgF "chromosome Y"
8. PgM "chromosome 1," PgM "chromosome 2," PgM "chromosome X"

As is obvious in these combinations, my brother and I could have inherited the same combination of these three chromosomes or a totally different one. Given that we are both male, all combinations containing chromosome X are excluded (3, 4, 6, 8), as we both have chromosome Y. Therefore, we must have one of the combinations 1, 2, 5, and 7. Some of them (e.g., 5 and 7) are quite similar, whereas others are more different (e.g., 1 and 7). An extreme case would be for my brother and I to have received almost entirely different chromosomes from our parents: I could have received all of the chromosomes of our grandmothers (all the PgM and MgM ones), whereas he could have received all of the chromosomes of our grandfathers (all the PgF and MgF ones), except for the PgF Y chromosome that both of us have received from our father. In another extreme case, we might have both received the 23 chromosomes of our paternal grandfather through our father, and 23 chromosomes of our maternal grandfather through our mother—but then we would be like twins, and this is not the case. The point here is that any two siblings inherit different combinations of maternal and paternal chromosomes, which in turn are—more or less—different combinations of the chromosomes of their grandparents (there may be epigenetic differences, but we can ignore them for now).

How are all of these different combinations of chromosomes produced? This is done through a type of cell division called meiosis or meiotic division (actually, this is the division of the cell nucleus, but I call it a cell division for the sake of simplicity). This division takes place only in the reproductive organs, and more specifically in the cells that will become spermatozoa (i.e., sperm) or ova (i.e., egg).[1] This is a division of cells with 23 pairs of chromosomes (called diploid) that results in new cells with 23 chromosomes each, actually one from each chromosome pair (these cells are called haploid). During meiosis, which consists of two division stages called the first and the second meiotic division or simply meiosis I and II, homologous chromosomes can be arranged in different ways in the various cells under division. As a result, an organism can produce different reproductive cells with different combinations of chromosomes. For instance, a diploid cell carrying chromosomes X, Y, 1_m, and 1_p ("m" standing for maternal and "p" for paternal, for that person's father and mother respectively), could produce haploid cells containing either X & 1_m and Y & 1_p, or X & 1_p and Y & 1_m. However, different cells could have different combinations of chromosomes. One combination of chromosomes could occur in some cells, whereas the other combination could occur in others. Therefore, some cells would produce spermatozoa with chromosomes X & 1_m and others with chromosomes Y & 1_p; other cells would produce spermatozoa with chromosomes Y & 1_m and others with chromosomes X & 1_p. But a man produces millions of spermatozoa at the same time. Therefore, a man could overall produce all the different kinds of spermatozoa: X & 1_m; Y & 1_p; X & 1_p; Y & 1_m (see figure 5.2).

The story is slightly different for women, as only a single ovum is released after meiosis. After the first meiotic division, an oocyte and a polar body are formed, and the latter is not further divided. Then, after the second meiotic division in that oocyte, the ovum and a second polar body are formed. However, which of the chromosomes will end up in the ovum and which ones will end up in the polar bodies is unpredictable and thus contingent *per se*. Once the chromosomes are allocated in these cells, the outcomes are contingent *upon* these divisions, as the offspring will come to have particular chromosomes but not others (see figure 5.3). For instance, if the initial cell had chromosomes $1_m, 1_p, X_m, X_p$ (1_m and X_m are the chromosomes of a woman's mother and 1_p and X_p are the chromosomes of a woman's father), then there could be several possible chromosome combinations in the ovum (1_m & X_m; $1_m, X_p$; 1_p & X_m; $1_p, X_p$), but only one of those will materialize. The offspring will receive only two of these chromosomes, and this depends on how the meiotic division that produces the ovum will take place. How chromosomes will be allocated to cells during this cell division is contingent *per se*, and what genetic material of maternal origin the offspring will have is contingent *upon* this division. This is why the moment of the alignment and the subsequent arrangement of chromosomes during the meiotic division that produces the ovum is a **turning point**.

Recall that we are using an abbreviated number of chromosomes (three) for simplicity's sake here. But each person has 23 chromosome pairs. So, the (mathematically) possible chromosome combinations in the reproductive cells of a human are many more. Because there are 23 pairs of chromosomes, there are 2^{23} mathematically possible combinations of those. Therefore, the mathematically possible number of reproductive cells

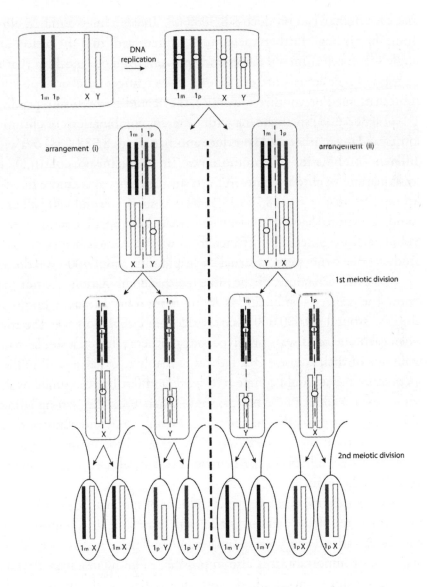

Figure 5.2. Meiosis is the cell division through which different spermatozoa with different chromosome combinations are produced. In the case of men, each individual may produce all different combinations of chromosomes, because arrangement (i) will occur in some cells and arrangement (ii) will occur in some others.

that each human can produce is 8,388,608. This is a huge number. Now, think this through further: each one of us stemmed from the fusion of a single sperm cell from our father and an ovum from our mother. That is, we were a single actual outcome among an even huger number of possible combinations. Put another way, my father could theoretically produce 2^{23} (or 8,388,608) spermatozoa with different combinations of chromosomes, and my mother could produce another 2^{23} (or 8,388,608) ova with different combinations of chromosomes. Therefore, the possible different combinations of chromosomes in the resulting embryos are more than 70 trillion ($2^{23} \times 2^{23} = 70,368,744,177,664$). As such, you and I and all of us could have more than 70 trillion (mathematically) possible siblings, who would each be genetically distinct from us. Now, each woman usually produces one ovum per menstrual cycle and thus about 400–500 during her whole life. Therefore, the possible *realized* combinations are not that many. The realized combinations are more for a fertile man, who might produce around 40,000,000 spermatozoa per ejaculation.[2] At the time when each one of us was conceived, our mothers released a single ovum from one of their ovaries. Our fathers probably released several million spermatozoa that would compete to reach and fertilize the single ovum. Some never made it to the uterus; some others took the "wrong" direction toward the ovary that had not released an ovum; and many of those others who took the "correct" way never reached the ovum. Therefore, from the initial millions of spermatozoa initially released, a few hundred made it to the ovum.

Now, which one of these several millions of spermatozoa will eventually fertilize the single ovum is entirely unpredictable: it is impossible to know in advance if any given spermatozoon will make it alive to meet the ovum; most important, it is also impossible to know in advance which one among those few hundred spermatozoa that reach the ovum will eventually make it through the ovum's membrane and fertilize it (note: the first spermatozoon that fertilizes the ovum initiates a reaction within the ovum that blocks the entry for the other spermatozoa). Once this

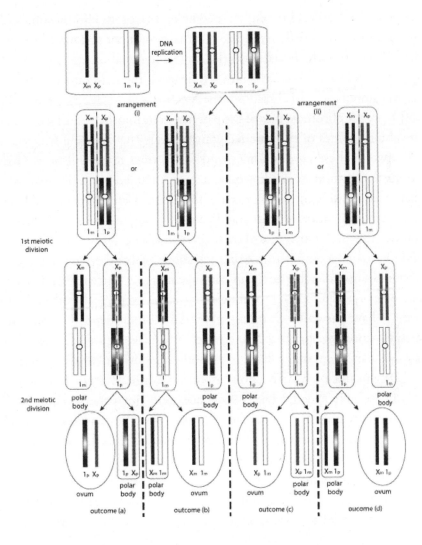

Figure 5.3. Meiosis is the cell division through which different ova with different chromosome combinations are produced. In the case of women, each individual may produce only one from the different combinations of chromosomes (outcomes [a]–[d]). Only one of the arrangements (i) and (ii), and only one of the outcomes (a)–(d) will occur. It must be noted that the second meiotic division in the ovum actually occurs after fertilization.

133

happens, the fertilized ovum has a complete set of 46 chromosomes (23 maternal ones already there, and 23 paternal ones in the spermatozoon that made it inside), which will form the genetic material of the resulting offspring. As already shown in figure 5.1, the available spermatozoa have different combinations of chromosomes: X & 1_m; Y & 1_m; X & 1_p; Y & 1_p (1_m and X are the chromosomes of the paternal grandmother, and 1_p and Y are the chromosomes of the paternal grandfather). The offspring will receive only two of these chromosomes, and which ones these will be are fixed during fertilization, depending on which are included in the spermatozoon that made it inside the ovum. Each of the millions spermatozoa will have a random combination of both grand-paternal and grand-maternal chromosomes. However, which one of these spermatozoa will fuse with the ovum and thus which combination of chromosomes will end up in it after fertilization is unpredictable and thus contingent *per se*. Once this is done, the outcome of development is contingent *upon* this event, as the offspring will come to have particular chromosomes and not others depending on which spermatozoon (see figure 5.2) and which ovum (see figure 5.3) fused. This is why the moment of fertilization is another important **turning point**.

Depending on which chromosomes we inherit from our parents, we also receive specific DNA sequences and not others. These DNA sequences are used as resources of information in the contexts of our cells for the production of molecules (proteins and others) that interact with one another, serve as signals for our cells, and eventually affect our development. Some of these sequences relate to the production of functional molecules that are implicated in the development of our biological traits, and we call them genes. Roughly put, we might say that, depending on which chromosomes we inherit from our parents, we also inherit particular versions of particular genes but not others. The reason for this is that for all genes there are different variant forms that may relate to a slightly different trait (how this happens is the topic of the next chapter). These different versions (different—usually slightly—DNA sequences) of the same genes are called alleles. Let us see how this works.

Let's imagine that the two X chromosomes that my mother has inherited from her parents are X_p and X_m, respectively. Let's also imagine three genes that are found on each of these chromosomes, which we can call g1, g2, and g3. My mother has two X chromosomes and therefore two g1 alleles, two g2 alleles, and two g3 alleles. Each of these alleles was received from each of her parents (my grandparents), and so we can call $g1_p$, $g2_p$, and $g3_p$ those alleles that my mother received from my grandfather, whereas we can call $g1_m$, $g2_m$, and $g3_m$ those alleles that my mother received from my grandmother (each of these combinations of alleles is called a haplotype). There are thus three pairs of alleles: $g1_p$ and $g1_m$; $g2_p$ and $g2_m$; and $g3_p$ and $g3_m$. Assuming that these genes are aligned on each of the X chromosomes, we can imagine that my mother's paternal chromosome X will contain alleles $g1_p$, $g2_p$, and $g3_p$, whereas my mother's maternal chromosome X will contain alleles $g1_m$, $g2_m$, and $g3_m$ (see figure 5.4). One reason that my brother and I might have different traits is that he received our mother's paternally derived X chromosome and I received our mother's maternally derived X chromosome, depending on which of the four outcomes (a), (b), (c), or (d) of figure 5.3 was actually realized in the production of the ovum from which we developed. However, you might think, even if my brother and I differ in the respective traits, each one of us has received a certain combination of alleles from our mother, so we should somehow resemble her or her parents, depending from whom this combination was derived. Nevertheless, there are other phenomena that may further increase variation.

One such phenomenon is called crossing over. What happens in this case is that during the meiotic division producing the ovum, the two X chromosomes exchange some parts and therefore some alleles (remember that each X chromosome consists of two sister chromatids). As a result of this exchange, new combinations of alleles may emerge (a similar phenomenon may occur during the meiotic divisions from which spermatozoa are produced). Thus, whereas my mother had in her X chromosomes two particular sets of alleles, $g1_p$, $g2_p$, $g3_p$ on one and $g1_m$, $g2_m$, $g3_m$

on the other, I may end up receiving a "new" chromosome X with a combination of alleles (a haplotype) that has not existed in any of my ancestors. In particular, whereas the haplotypes of my mother were $g1_p$, $g2_p$, $g3_p$ and $g1_m$, $g2_m$, $g3_m$, I could have ended up having a novel haplotype $g1_m$, $g2_m$, $g3_p$ as a result of crossing over (see figure 5.4). Thus, even though I have inherited the alleles that my mother had, I have nevertheless inherited a combination of those that neither she nor her parents had. In this case, I would actually carry alleles from my grandfather and my grandmother together on the same chromosome!

In this sense, the moment of crossing over is another **turning point**. Whether or not crossing over will occur is totally unpredictable, and thus it is contingent *per se*. Where exactly crossing over will take place is also unpredictable, although we are aware that there exist certain hot spots in the human genome. This means that crossing over is more probable is some areas of the genome than in others.[3] Nevertheless, being more or less probable is different from being predictable, as even if crossing over is probable in a certain area of the genome, it may never occur. Now, once it occurs, the new haplotypes that are subsequently produced may have an impact on our traits, and in this sense these traits are contingent *upon* the crossing over event. Of course, crossing over can further raise the possible variation among the mathematically possible offspring of two people. This means that until the moment of fertilization, after which the spermatozoon and the ovum will fuse, the segregation of homologous chromosomes and their recombination brought about by crossing over can produce novel combinations of alleles that are unique to us.

This is why one cannot tell in advance what features one's children will have. It is totally unpredictable which combination of chromosomes will exist in the spermatozoon and the ovum from the fusion of which a new person will develop. It is also impossible to predict which new combinations of alleles may arise on the chromosomes that a person inherits from the parents. This explains why siblings can be very different.[4] This also suggests that, even if you think that you were somehow meant to be born, ultimately you were not. All of us are contingent outcomes: contingent *upon* the meiotic divisions in our parents' reproductive organs, which themselves were contingent *per se*. This has important implications for our sense of identity. In a way, there is a genetic lottery at work, which produces our genetic constitution. So the next time you listen to a parent telling with affection his/her baby "we have been waiting for you to come," you can explain to this person that he/she was expecting for a baby to come, but not necessarily this particular one. Which combinations of genes this baby has is contingent upon the shuffling of chromosomes and alleles during meiosis in the reproductive organs of its parents.

Why does this matter? Biologically speaking, each one of us is as unique as one can be. Our biological characteristics as well as our talents and behaviors are the outcomes of the interactions between our genome and our environment. These interactions are unique for each one of us, because we each have a unique, and previously unpredictable, combination of the DNA of our ancestors. Therefore, at the biological level we have a unique genetic heritage that has most likely never existed in the past and that will most likely never exist again.[5] Why is this? Because no matter how many people have lived on Earth, if the mathematically possible offspring of each couple is over 70 trillion, you can imagine that it is entirely improbable for the same combination to exist twice. But this is not the most important take-home message. Rather, what you should keep in mind from this chapter is that our conception comprises important turning points relating to the meiotic divisions in the reproductive cells of our parents. This means that none of us was *a priori* meant to come to this life. That is, none of us was destined to be.

Another important issue relates to our sense of identity. We often perceive similarities between ourselves and not only our parents, but also our uncles, aunts, and grandparents. When it comes to biological characteristics, there is definitely a component of genetic inheritance that relates to any similarities observed. However, it is also important for us to recognize not only how much of our DNA is common with our relatives but also

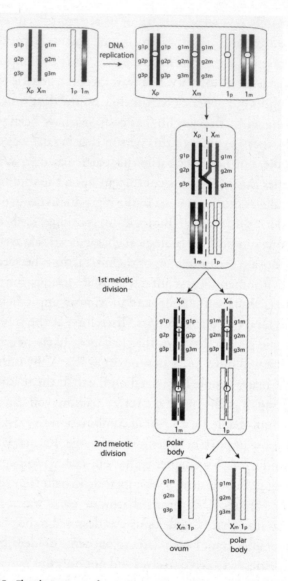

Figure 5.4. The phenomenon of crossing over can produce new combinations of alleles in the ovum (and eventually in the offspring) that existed neither in the mother nor in her parents. It must be noted that the second meiotic division in the ovum actually occurs after fertilization.

how much of it is not common. The common and non-common DNA can in part account for our similarities and differences. To better understand this, consider the family tree depicted in figure 5.5. Males are represented with rectangles and females with circles; spouses are connected with a horizontal line; siblings are connected with a horizontal line above them that also connects them to their parents. We can identify each member of the family by referring to the generation to which this person belongs, (I), (II), or (III), and to each person by his or her number, from left to right. Therefore, for example, I1 and I2 are spouses and also the parents of II2 and II3. So, the question here is: how much of their DNA do the members of this family share? Remember that what we inherit from our parents, and through them from our ancestors, are chromosomes. Chromosomes differ in two important aspects: (1) the sequence of DNA included in them, the inheritance of which is described as genetic inheritance; and (2) particular marks in the histones with which DNA is packed, the inheritance of which is described as epigenetic inheritance.

Let us start with DNA. Any parent will always share about half of his or her DNA with any of his or her children. As already explained above, because of the way in which chromosomes (and therefore the DNA included in them) are segregated and assorted during meiosis, each one of us receives 50 percent of our DNA from each one of our parents. This is an approximation, as there can be some variations. For instance, the X chromosome that men inherit from their mothers is larger than the Y chromosome that they inherit from their fathers, but this does not significantly affect that fact that their genetic contribution is about 50 percent. This is the case, for example, in the family tree in figure 5.5 between the woman II3 and her daughter III3. It can never be the case that a person has received 20 percent of his or her DNA from one parent and 80 percent from the other parent. The only other case in which we can be certain about what is going on is that of monozygotic twins, that is, siblings who emerged from the same embryo. This happened because at some point during the development of that embryo, usually before implantation in the uterus, its cells were separated into two groups, each of which developed to an individual.[6] Because they emerged from the same fertilized ovum, these siblings have the same genetic material; that is, they should have about 100 percent the same DNA. When we are looking at the genetic similarity among family members, there are only, therefore, two cases about which we can be certain: (1) children receive about 50 percent of their DNA from each parent, and (2) monozygotic twins share about 100 percent of their DNA. To account for these certainties, the percentages are highlighted in black boxes in figure 5.5. (Note: in both cases I intentionally described the amount shared as "about 50 percent" or "about 100 percent" because, as I explain in the next chapter, it is possible for new variations to emerge along the way.)

In all other relations presented in figure 5.5, the percentages of common DNA are averages. This means that any two siblings who are not monozygotic twins might share more or less than 50 percent of their DNA—50 percent would be the average. Let us see why. An extreme case would be for two siblings to receive exactly the same 23 chromosomes from each of their parents. However, given the numerous possible combinations, this is extremely unlikely to happen. But other cases are possible. For instance, two brothers (III1 and III2) could inherit exactly the same 23 chromosomes from their mother (e.g., the 23 chromosomes of their grandfather I1 that their mother, II2, has), as well as exactly the same twelve chromosomes and eleven different chromosomes from their father, II1 (e.g., one brother got all chromosomes from his paternal grandfather whereas the other got half of these and half of the chromosomes from his paternal grandmother). These two brothers would therefore have about 75 percent of their DNA in common. Another example in the opposite direction would be for those two brothers to have inherited totally different chromosomes from each of their parents, with the exception of their father's Y chromosome. In such a case, one of the brothers could have inherited all the grand-paternal chromosomes through his parents, whereas the other could have inherited all the grand-maternal chromosomes through his parents. In this case, the two brothers would have very different DNA, with only their father's Y chromosome in common between them (which would amount for less than 5 percent of their total DNA). In the same sense, the girl III3 would share on average 25 percent of her DNA with her aunt II2 and her grandfather I1. Therefore, the saying—which is very popular to my country of origin—according to which grandparents say "my child's child is twice my child" is not accurate, genetically speaking. It is actually on average half their child, or 25 percent of themselves.

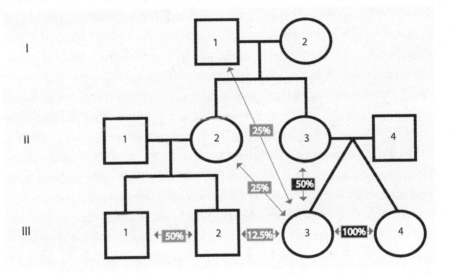

Figure 5.5. Monozygotic twins share about 100 percent of their DNA; parents and offspring share about 50 percent of their DNA. In all other cases, the estimated percentages are averages, as it all depends on the outcomes of meiotic divisions.

So far we have discussed genetic inheritance; the second point to consider relates to epigenetic inheritance. As already mentioned, DNA in cells is not "naked," but it is packed with proteins called histones thus forming chromatin, which during cell division becomes further compacted and forms chromosomes. In recent years, however, researchers have come to understand that chromatin is more than just a means of packaging DNA in the nuclei of cells. Several phenomena bring about modifications in chromatin, which do not affect the sequence of genes but which may affect their expression, that is, whether a molecule with a functional role (protein or RNA—ribonucleic acid, a "cousin" of DNA) will be produced. Some of these modifications can be stable and be transferred across generations, that is, be inherited. As these happen not within DNA but upon it, this type of inheritance is described as epigenetic. Generally speaking, we might say that "epigenetics" refers to all interactions between DNA and the molecules around it that eventually have an impact on gene expression. In particular, specific modifications of either DNA or histone proteins can enhance or block the expression of genes. One example of such modifications is called DNA methylation. Simply put, the addition of a very small chemical group (methyl group) to cytosines (C) found "before" a gene may block the expression of that gene. When Cs "before" the gene are highly methylated, some proteins can bind there and prevent other proteins from "reading" the gene and "expressing" it, that is, produce whatever functional molecule it can produce. The methylation pattern of a DNA molecule can be preserved after DNA replication (which takes place before cell division) and thus actually can be transferred to new cells. Methyl groups can also be added to the histones surrounding DNA. In some cases, methylation of a certain histone seems to enhance DNA methylation, whereas in others it seems to stop it. Other molecules, such as acetyl groups, can be added to histones, and the result of this can be the increased expression of genes (this phenomenon is described as histone acetylation).[7]

Details notwithstanding, the important point is that it does not only matter which genes we inherit from our parents, but also which epigenetic modifications go with them, as these might affect whether or not these genes could be used by our cells. This simply means that we might have genes that are not used by our cells because their expression is blocked. This research is quite recent and so there is a lot that we still do not know. Epigenetic modifications are generally stable in the somatic cells of mammals. However, all or most (this is not entirely clear yet) of them seem to be erased and created anew on a genome-wide scale in the primordial reproductive cells, and in the early embryo after fertilization and before the blastocyst stage (when the embryo can implant on the uterus). This process is described as epigenetic reprogramming, and as a result the epigenetic marks are not transmitted unchanged across generations, but are each time created anew. It is not clear why this is happening, but as epigenetic marks relate to the ability of a cell to proliferate and produce one or many types of tissues, it seems that the erasure of the epigenetic marks after fertilization is necessary in order for the cells of the early embryo to become able to divide, differentiate, and produce the various tissues of the body. Now, despite these processes, it has been shown that some epigenetic marks may escape reprogramming at both phases. Therefore, it is possible for transgenerational, epigenetic inheritance to occur, but there is a still lot to learn about these processes.[8]

With the examples discussed in this chapter, it should by now be clear that our own existence was not inevitable, but totally contingent *per se* and also *upon* critical events that involved our parents' reproductive cells. The potential for variability that these cells have is enormous when one considers the various theoretically possible combinations of chromosomes within these cells. Not only the emergence of each one of us as a zygote is unpredictable, but we are also just one of trillions of theoretically possible offspring that our parents could produce. This highlights our uniqueness. No matter how many children our parents could have, it is extremely unlikely for a sibling even closely resembling me or you to be born. We all share 50 percent of our DNA (on average) with our siblings, and we might actually share more or less than that. But there will always be differences because it is highly improbable, and practically impossible, for the same 1 in 2^{23} (or 1 in 8,388,608) spermatozoa with different combinations of chromosomes and the same 1 in 2^{23} (or 1 in 8,388,608) ova with different combinations of chromosomes to emerge twice. Our existence and our own genetic constitution is highly contingent and thus unique. This becomes even clearer when we consider how our biological characteristics develop.

Chapter 6

THE APPLE MIGHT FALL
FAR FROM THE TREE

Individuals tend to physically resemble one or both of their parents. Therefore, if someone asked you whether a blue-eyed woman or a brown-eyed woman is the mother of a blue-eyed girl, you would be tempted to suggest that the blue-eyed woman is the mother. This is indeed a likely accurate response, although it is also possible than the brown-eyed woman is the mother and the girl's eyes resemble those of her father. It would also be possible that none of the parents has blue eyes and that the girl has come to have the blue eyes due to some ancestral genetic influence, and a combination of genes that she inherited from her parents. So, here are some different possible explanations: (1) the girl has blue eyes because her mother also has blue eyes; (2) the girl has blue eyes, even though her mother has brown eyes, because the girl's father has blue eyes; or (3) the girl has blue eyes, even though both of her parents have brown eyes, because one of her ancestors (a grandparent or an even more remote ancestor) also had blue eyes. There are two important points to note here. The first is that our biological characteristics do not necessarily reflect our genetic potential and predisposition, as a person having brown eyes might also carry and transmit DNA that, when combined with other DNA in the offspring, might result in the latter having blue eyes. The second is that, therefore, even though explanation (1) may seem more likely than explanation (2), which in turn may seem more likely than explanation (3), this is not necessarily the case. Indeed, all of these explanations are possible.

It is often thought that our biological characteristics are predetermined in our DNA, and that they will invariably develop as we grow up. However, this is not accurate. Besides DNA, several other factors that affect development may have an influence on our characteristics. The human genome certainly affects human development; however, the outcome of development also depends on the complex interactions within cells between the products of genome expression and molecules from the intra- and intercellular environment. Outcomes that depart from what we consider as "normal" are rare, but their occurrence shows that it is critical events during development that shape these outcomes. Elsewhere I have described in detail how human traits such as eye color, height, and behavior develop.[1] In this chapter, I provide a general overview of these processes, highlighting some crucial moments during development. In the previous chapter I explained how each one of us comes to have a previously unpredictable combination of chromosomes and therefore of genes and other DNA sequences. But simply having these genes and other DNA sequences does not invariably predetermine the outcome of our development. There is also the question of how these genes and other DNA sequences are used by our cells.

So the question arises: Do some traits, skills, or talents run in families? Indeed, this seems to be the case. For instance, George H. W. Bush the elder and his son George W. Bush were both elected presidents of the United States. Kirk Douglas and his son Michael became famous Hollywood stars. People often look at cases like these and think that the success or failure of children was to be expected, given the success or failure of their parents. As it is commonly said: "The apple does not fall far from the tree." No matter whether the cause of this is genetic or environmental, it seems natural. And it seems even more natural that children will resemble their parents in their biological characteristics. Kirk and Michael Douglas, as well George Bush the elder and the younger, certainly look similar; "like father, like son" people would say. And if you look around, in your family or among your friends, you will certainly find striking similarities between relatives. Among the various traits, there are some that we consider especially "inheritable," such as the color of our eyes. My wife, her father, and his father all have the same green eyes. In contrast, my maternal grandmother, my mother, and I all have the same (boring) brown eyes. It is heredity after all, isn't it?

The problem is that we pay a lot of attention to characteristics that run in families, and we forget numerous other cases in which this does not happen. How many successful politicians and actors had parents who were not equally successful? Many of them! How many of the children of these successful politicians and actors followed the steps of their parents? Not many of them! Furthermore, how many children do not resemble their parents at all, or—even better—have characteristics that neither of their parents has? If you start thinking carefully about this, you will realize that this happens in a lot more cases than perhaps you had previously thought. Why does it happen? The first reason has already been explained in the previous chapter: each one of us inherits a unique combination of maternal and paternal chromosomes (and in turn grand-maternal and grand-paternal ones). These chromosomes may also contain the same combinations of alleles as our ancestors, or different ones that are the result of the recombination of chromosomes. As explained in the previous chapter, until the moment that the reproductive cells of our parents fused, different combinations of chromosomes and alleles on the same chromosome were possible.

Let us consider a concrete example: the color of our eyes. The variations we observe in human eye color is due to variation in the coloration of the iris, which is the circle around the pupil through which light enters the eye. The outer tissue layer of the iris includes melanocytes, that is, cells that store melanin in organelles called melanosomes. It is the amount of melanin and the number of melanosomes that mostly affect what color the iris will come to have. When white light enters the iris, it can absorb or reflect different wavelengths, thus giving rise to the common eye colors and their variations that we observe. Roughly put, it can be said that blue eyes contain minimal melanin levels and melanosome numbers; green-hazel eyes have moderate melanin levels and melanosome numbers; and brown eyes have high melanin levels and melanosome numbers (see figure 6.1). This means that the eye color that we perceive is due to not some colored substance found inside the eye but the amount of melanin available there, as well as the thickness, constitution, and density of the iris itself. This is why the variation in the eye color of humans is enormous (just observe carefully the eyes of people around you, but first alert them that you are doing this out of scientific curiosity). Therefore, simplistic descriptions of eye color inheritance, such as those taught at schools (i.e., that a "dominant" allele B controls brown color, and a "recessive" allele b controls blue color) fail to account for the variation observed. According to such narratives, parents with brown eyes can have children with blue eyes, but parents with blue eyes cannot have children with brown eyes. There two problems with such teachings: (1) eye colors are not of the "either-or" type (e.g., either brown or blue), but instead there exists a continuum of colors among them; (2) the alleles do not "control" the eye color but are implicated in its development. Like all our biological characteristics, eye color—or, better, iris coloration—is the outcome of developmental processes.[2]

Even though we often think of eye color and other biological characteristics as inherited ones, what is inherited is not the characteristic itself but a developmental predisposition that will be realized depending on which versions of genes and other DNA sequences will come together in an individual, and when and how they will be expressed in the iris cells. I have dark-brown eyes and my wife has green eyes, but none of our children has the same eye color with either of us. Instead, our daughter has brown eyes that are lighter than my own, whereas our son has a mixture of brown and green color in his iris. What our children inherited from my wife and I are certain DNA sequences; their combined expression during development resulted in the eye color that they now have. The point I want to make here is that the eye color that our children, and in fact all of us, have is a matter of contingencies and not at all predetermined. Let me explain why.

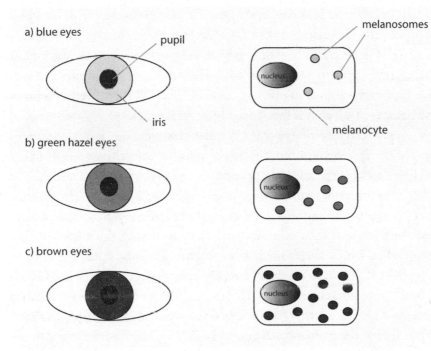

Figure 6.1. The various colors of the iris depend on the number of melanosomes within each melanocyte, as well as on the amount of melanin within each melanosome. Blue eyes have few melanosomes with a small amount of melanin within them, whereas brown eyes have many melanosomes with a big amount of melanin within them. The three common colors are blue, green-hazel and brown, but in reality a continuum of colors exists among them (just do a search on the web using "eye color variation," and you will see).

We currently do not know exactly how eye color develops, but associations have been found between eye color and particular genes. The most notable associations have been found between particular eye colors and variants of two genes on chromosome 15, the *OCA2* and *HERC2* genes (it is believed that the latter affects the expression of the former). Here is a model proposed about how the *OCA2* gene relates to iris color: when this gene is expressed, the OCA2 protein is produced, which promotes the maturation of the melanosomes, and eventually a fully pigmented brown iris results. This happens when a *T* allele of gene *HERC2*, which regulates the expression of the *OCA2* gene, exists. In contrast, when a *C* allele of gene *HERC2* exists, the OCA2 protein is not produced and therefore there is no production of melanosomes, which results in a blue iris. However, variants in other genes also seem to be involved, so human eye color is a trait that is influenced by several genes.[3]

The inheritance of eye color is quite complex, but let us consider a simple example here in order to make sense of how things work. Let us assume that *O* is the allele that produces the OCA2 protein and *N* is an allele that does not; similarly, let us assume that *T* is the *HERC2* allele that enhances the expression of the *OCA2* gene, whereas the *C* allele does not. We can thus assume that a woman having the alleles *NNTC* (remember, which alleles she has is her genotype) and a man with genotype *ONCC* will have blue eyes because they do not produce the OCA2 protein, for different reasons. The man with genotype *ONCC* has the necessary *OCA2* allele (*O*), but it is not expressed because he also lacks the *T* alleles; the woman with genotype *NNTC* does not have the necessary *OCA2* allele. If these two persons decided to start a family and have offspring, what eye color would you expect them to have? Many people would say that parents with blue eyes can only have children with blue eyes. Let us follow the meiotic divisions in order to see what kind of spermatozoa and ova could be produced, and therefore what kind of combinations of alleles the offspring might receive from their parents. The father could produce different kinds of spermatozoa with either the *OC* or the *NC* alleles, whereas the mother could produce either an ovum with the *NT* alleles or an ovum with the *NC* alleles. The possible combinations are shown in figure 6.2. It becomes evident that there is a probability of 75 percent that their child would have blue eyes as well, if the child had any of these combinations of alleles: *NNTC* (no *OCA2* allele), *NNCC* (no *OCA2* allele), *ONCC* (*OCA2* allele exists but is not expressed). However, there also would be a probability of 25 percent that the child would have the alleles *ONTC*, and carry both the *OCA2* and the *HERC2* alleles that result in brown eye color. Two blue-eyed parents would thus have a child with brown eyes.[4] As you can see, this is an example of how children can exhibit features that their parents lack.

The above shows once again that the segregation of chromosomes during meiosis is a **turning point**. Which of the two homologous chromosomes, that is, the two similar chromosomes that are considered to form a pair and that are each derived from one of our parents, with the alleles *NT* or *NC* will be included in the ovum is contingent *per se*, as one cannot predict in advance which chromosomes will end up there and which ones will end up in either of the polar bodies (see figure 5.2). Then, the subsequent development of the child is contingent *upon* this event, because its phenotype will depend on the alleles included in the ovum (*NT* or *NC*) as well as those in the spermatozoon that will fuse with it (*OC* or *NC*). (Note also that, as already explained in the previous chapter, the event of fertilization is itself a turning point.) To assess the importance of the outcome in the meiotic division in the ovum, we should remember that a woman typically produces only one ovum (with a single chromosome for each pair) per menstrual cycle, whereas a man can produce millions of spermatozoa with all different chromosomes being available at the moment of fertilization. Therefore, how chromosomes segregate during the meiotic divisions in the ovary is critical because only one from each pair of homologous chromosomes will be included in the ovum and thus be passed on to the offspring. The respective phenomenon taking place in the testes is not that critical, because millions of spermatozoa are produced each time; this means that all chromosomes will be available for fertilization, albeit not in all possible combinations, and might be passed on to the offspring at the moment of fertilization. In that case, it is which one will make it into the ovum that matters.

So far, so good. The combination of chromosomes and alleles that each one of us has is the outcome of contingencies. But is it only this combination that matters, or does the DNA included in chromosomes also change after fertilization? Indeed, it is possible for changes in our

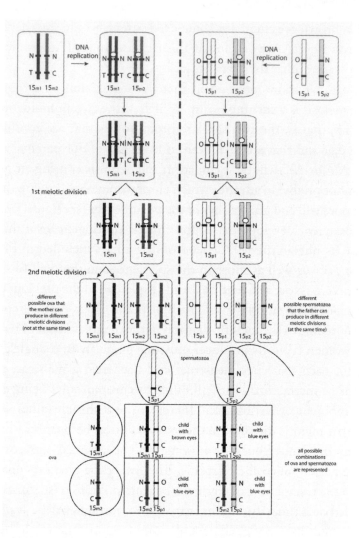

Figure 6.2. How two blue-eyed parents can have a child with brown eyes. The parents can produce different reproductive cells with different alleles, and which ones a child will come to have will depend on which ones will be included in the reproductive cells that fuse during fertilization. It must be noted that the mother cannot produce all of these ova at the same time as the father can (see figures 5.2 and 5.3 in the previous chapter). It must also be noted that the second meiotic division in the ovum actually occurs after fertilization, but to avoid confusion here these are presented as independent events.

DNA to take place after fertilization. These changes are usually described as mutations; this is a term with a negative connotation, nevertheless it means only "change," and change is not always bad. DNA replication usually takes place without mistakes; but even if mistakes occur, cells have proteins that "proofread" and "correct" them. As a result, two identical DNA molecules are expected to emerge from the replication of DNA (figure 6.3a). However, it is possible that mistakes, mutations, exist in the end. As mentioned above, mistakes can occur during DNA replication, for example. In the simplest case, there can be a change in a single nucleotide. In some cases, a base might end up pairing by mistake with a base different than the one it usually pairs with. For example, base G usually occurs in a certain form that pairs with C. However, it can sometimes change to another less stable one (G*) that can pair with T. As a result, during DNA replication it is a T and not a C that is inserted opposite a G*. In the next round of replication, T will be paired with A, thus producing a change in the sequence of DNA, as the pair in that position will thereafter be A-T (see figure 6.3b) and not G-C, as it was in the initial molecule (6.3a). Mutations can also take place due to environmental factors that directly affect the structure of DNA, such as ultraviolet (UV) light. This causes two Ts next to each other to join, and when these are later replaced it is possible that other bases (e.g., two As) are put into their place (6.3c). These kinds of changes are the simplest ones possible, and other extensive mutations that involve larger DNA sequences or whole chromosome parts are also possible. But such simple changes may have large effects, as I explain in this and the following chapter, and it is interesting to see what happens.

Therefore, once we are conceived and once we receive whatever combinations of chromosomes and genes from our parents, the story is not over. In fact, because of mutations that may take place during development, we may end up having characteristics that our ancestors did not have. It has actually been estimated that each one of us can carry between 72 and 138 novel mutations.[5] This number is small, given that we have in

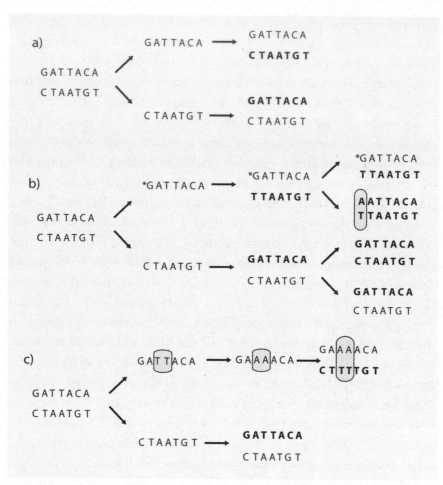

Figure 6.3. (a) The replication of DNA results into two identical DNA molecules. The newly synthesized strands are highlighted in bold. Mutations, circled here, can occur during the process of DNA replication and eventually result in new DNA sequences. This can happen by mistake, (b), or result from environmental factors, (c).

each of our cells about 6 billion nucleotides. This is the reason why each of the various mutations that each one of us will have is a **turning point**. Whether or not a mutation such as those depicted in figure 6.3 will occur is unpredictable and contingent *per se*, depending on the probabilities of a mistake to occur during DNA replication either by accident or because of environmental factors. Note, though, that the subsequent events will be contingent *upon* that mutation. A change like the one depicted in figure 6.3(b) might have no impact at all—for example, this change might occur on a DNA site that has no function at all; it might also occur within a gene, but the sequence of the corresponding (RNA or protein) molecule might not change at all or change at a part of the molecule that does not affect its function. However, some mutations can bring about significant effects with important implications, such as changing an allele related to a characteristic (like eye color) but most importantly diseases (as explained in the next chapter). This depends on how the changes in DNA might affect the resulting functional molecule (RNA or protein).[6]

Let us use a simple analogy to illustrate this, without getting into the details of gene expression. Imagine that you are reading the recipe for a cake, which includes the following ingredients: eggs, flour, milk, butter, and sugar. Imagine that this is a sequence of DNA and that the cell can draw upon this resource to construct a product: a cake. The recipe is just a generative plan about which ingredients are required and about how they should be mixed together in order to make a cake. This is quite similar to the kind of information that DNA contains. Beyond that, how and when the cells use this plan depends on local signals from inside and outside the cells. Assuming that all of these ingredients are necessary for the cake, here are some different single-letter changes that can take place:

(1) eggs; flour; milk; butter; sugar → eggs; flour; milk; butter; sugar
(2) eggs; flour; milk; butter; sugar → eggs; flour; milk; batter; sugar
(3) eggs; flour; milk; butter; sugar → eggs; flour; silk; butter; sugar
(4) eggs; flour; milk; butter; sugar → eggs; floor; milk; butter; sugar

Changes like the one in 2 might not be very significant. Batter is itself a mixture like the one we would make if we mixed the ingredients in (1), so it might work. But if you tried to add silk instead of milk in the mixture, things might go terribly wrong for the recipe. And if one asked you to add a floor instead of flour, you would most likely not know what to do. This very simplistic example shows that there are some changes that may not change the product significantly (2), that some changes may change it significantly (3), and that some changes might make it impossible to produce (4). Shifting back to genetics, this means that mutations that might occur in an individual might result in the anticipated eye color or in an eye color different than what would be expected, or in no eye color at all (as in the case of albinism).

However, it is not only mutations that can bring about changes. Developmental processes can produce different outcomes even without changes at the genetic level. One interesting example that shows the impact of developmental processes on phenotypes is heterochromia, the condition in which a person has eyes of two different colors. When the irises of the eyes are entirely different from each other, then the condition is described as complete heterochromia. When it is only part of one iris that is a different color than the rest of it, the condition is called partial heterochromia. Finally, when a person has an inner ring around the pupil of the eye that is a different color than the outer area of the iris, the condition is called central heterochromia. In general, heterochromia can be acquired because of eye injury or eye bleeding, or it can be the result of particular syndromes. When it exists at birth, it is rarely related to syndromes, and in most cases, there is no detectable underlying abnormality.[7] It also seems that in most cases there is no family history. This condition is rather rare, but it is an interesting example of changes that can take place during development. In a study of 7,000 children in the United States, 47 cases of heterochromia were found (frequency: 0.671 percent or 6.71 in 1,000). Another study with 25,346 individuals in Austria found 65 cases of heterochromia (frequency: 0.256 percent or 2.56 in 1,000). Whereas the observed frequency was lower in this second study, it was noted that five out of six of all cases of heterochromia were found in children and adolescents (aged two to nineteen years). Finally, it was also found to be more than twice as frequent in women as in men.[8] Whatever the case, the condition is frequent enough to be observed in famous movie stars and athletes and has merited mention in the news.[9]

The point I want to emphasize here is that biological traits can be highly variable and are in no way predetermined in our genes. The processes of development, and whatever takes place during them, also have an important impact. Therefore, we should not think that our adult traits are fixed at the moment of our conception, depending on which chromosomes and genes we inherit from our parents. Rather, how these will be "used" by cells, makes a difference. What is more important is that these can even be modified along the way. This is most clearly illustrated in the case of monozygotic twins, who at the moment of their conception have the same DNA (because they emerge from the same initial fertilized ovum, or zygote, hence the term monozygotic). Several examples of genetic differences between monozygotic twins have been found and these often relate to the number of copies of particular DNA segments (usually described as copy-number variations, or CNVs). CNVs can exist between the cells of an individual or between individuals, and they are due to mistakes in crossing over that result in insertions, deletions, or duplications of chromosome fragments. For instance, in a study of 1,097 monozygotic twin pairs, a total of 153 CNVs between twins were found.[10] Whereas more work is required to arrive at a finer level of analysis, in order to see whether CNVs contribute to phenotypic variation, it is important to realize that changes after one's conception are possible.

However, besides the DNA and genome, there is an additional level with a potentially critical influence on traits: the epigenome. As already described in chapter 5, epigenetic changes are changes "upon" DNA itself, either directly or via other modifications on the histone proteins to which it is linked. It has long been shown that epigenetic differences can account for observable differences in traits between monozygotic twins. For instance, a study with eighty Spanish monozygotic twins showed that although they were indistinguishable during their early life in terms of DNA methylation and histone acetylation, as they grew older the differences in the distribution of these marks across the genome became remarkable.[11] Interestingly, these differences were more significant between twins who had different lifestyles and had spent less of their lives together. These twins had the most-different gene expression profiles; between fifty-year-old twins there were four times as many differentially expressed genes as in the three-year-old ones (whose expression patterns were almost identical). These findings suggest that even people with the same DNA may develop significantly different traits. The differences in the epigenetic modifications could be due to the long-term influence of external factors such as smoking, diet, or physical activity. However, it could also be due to changes in the transmission of epigenetic marks through successive cell divisions or due to their maintenance in differentiated cells, in other words, due to a process of epigenetic "drift" (in the sense of random sampling) associated with the process of aging.

Let us consider a particular example. Perhaps you have wondered so far if it makes any difference that males have a X chromosome and a Y chromosome, whereas females have two X chromosomes. This might make a big difference if it weren't for a specific mechanism that inactivates one of the X chromosomes in females. As a result, both males and females have one active X chromosome. This means that although a girl will receive a paternal and a maternal X chromosome, the genes in only one of them will actually be expressed. This is due to complex mechanisms of regulation and X chromosome inactivation that take place early in development, when the embryo consists of a limited number of cells as a blastocyst (the form with which the human embryo implants on the uterus of the mother). This process has been studied in detail in mice, where the process of gene inactivation on the X chromosome is affected by a DNA region known as the X inactivation center, from which RNA molecules are produced that block the expression of X-linked genes. A major feature of the inactivated X chromosomes is the increased level of DNA methylation and of histone modifications that block gene expression. What is more important is that this process is rather random, which means that each cell may end up having either the paternal or the maternal X chromosome being inactivated. Thus, females are in fact mosaics that consist of cells with different inactivated X chromosomes and different patterns of expression of X-linked genes.[12] The classical illustration of this phenomenon is calico cats, in which the females have a mosaic of black and ginger spots because a different X chromosome is inactivated in each group of cells.

Another interesting case of this kind is genomic imprinting. This is the phenomenon in which one of the alleles of the same gene, either the one on the paternal or the one on the maternal chromosome of a certain pair, is "switched off." This is due to epigenetic marks, such as DNA methylation, on particular regions of the one but not of the other chromosome. As a result of this, only the allele lacking the epigenetic marks that maintain the "switching off" is expressed. Under certain conditions the epigenetic imprinting mechanism may not work well and imprinting may be lost, thus resulting in a normally silent allele being activated or a normally active allele being silenced. An example involves two human genes: *IGF2* and *H19*. The *IGF2* gene is a key factor in human growth and development, and relates to the production of a protein called insulin-like growth factor 2, which promotes the proliferation and differentiation of cells in various tissues. The *H19* gene relates to the production of an RNA molecule that has important regulatory roles. In this case, there exist certain regions on the paternal and the maternal chromosomes 6 that are differentially methylated (and thus are called differentially methylated regions, or DMRs). One of them is found between the *IGF2* and *H19* genes, and it influences the expression of both (so let's call it *IGF2/H19* DMR), and the other is found within the *IGF2* gene (for simplicity, let's call it *IGF2* DMR). The *IGF2/H19* DMR is methylated on the paternal chromosome, whereas the *IGF2* DMR is methylated on the maternal chromosome. As a result, the *H19* paternal allele and the *IGF2* maternal allele are normally inactive.[13] There are some possible changes in the methylation status of these two DMRs that eventually result in the expression of both the maternal and the paternal alleles. Let us see why and what the related implications are.

A striking example of how epigenetic changes can impact adult characteristics is that of the so-called Dutch famine. During the winter of 1944–1945, while the Netherlands was under German occupation, there was shortage of food in the western part of the country due to a German-imposed food embargo. Although the southern part of the country was already liberated by the Allies, this was not yet the case for the northern part. To support the Allies, the exiled Dutch government organized a strike of the national railways in an attempt to delay the transfer of German soldiers. The Germans responded by putting an embargo on all food transports. Even though the embargo was partially lifted in November 1944 by allowing transportation of food across water canals, an unusually severe winter caused all canals to freeze and so food transportation became impossible. As a result, the food stocks diminished. Between December 1944 and April 1945, people in the affected region, including pregnant women, received around 400–800 calories per day, and were thus seriously undernourished.[14] A study looked at 2,414 people born between November 1943 and February 1947 in Amsterdam, for whom detailed birth records were available.[15] Those people whose mothers had received less than 1,000 calories per day during the first, second, or third trimester were considered exposed to the famine, whereas those who never received less than 1,000 calories per day were considered unexposed to the famine. The researchers found that the long-term consequences of exposure to famine varied depending on the trimester of the gestation during which the exposure to famine occurred. People born to mothers exposed to famine in the first trimester of pregnancy were found to have more coronary heart disease, a more atherogenic lipid profile, disturbed blood coagulation, increased stress responsiveness, more obesity, and—for women—increased risk of breast cancer. People born to mothers who were exposed to famine in the second trimester had more microalbuminuria and obstructive airways disease. Finally, all people born to mothers who were exposed to the famine at any stage had reduced glucose tolerance and increased insulin levels. But what does epigenetics have to do with these outcomes? Well, it seems that epigenetic differences underlie these different outcomes.

A study of people affected by the Dutch famine aimed at investigating whether prenatal exposure to that famine is associated with differences in methylation of the *IGF2* gene.[16] For this purpose, 60 individuals who were conceived during the famine and 62 individuals who were born right after it were selected to participate in the study. The levels of *IGF2* DMR methylation were then compared between these individuals and their siblings. In the first group (exposed to famine during early pregnancy), *IGF2* DMR methylation was found to be lower than their siblings for 43 out of 60 individuals. In contrast, no difference was found between the individuals of the second group and their siblings. It was concluded that those people who were exposed to the famine around the time of their conception and early in their mother's pregnancy had overall lower methylation of the *IGF2* DMR sixty years later. In contrast, those people who were exposed to famine during the later stages of pregnancy did not have this feature. This means that their mothers' exposure to the famine during early pregnancy had long-lasting implications for the epigenome of these people.

The Dutch famine story is an example of how environment conditions during our early, prenatal development can cause epigenetic changes that persist through life. There now exist several such examples of how environmental conditions can bring about epigenetic changes that eventually affect development.[17] There is a lot more to learn, and it seems very likely that during the next years we will better understand the interactions between our genome and the environment. But it is already clear that the conditions of fetal developmental that might bring about epigenetic changes are a **turning point**. Whether these conditions will exist at all is previously unpredictable and thus contingent *per se*, as the Dutch famine was the outcome of a combination of various factors explained above. Of course, famine is not a surprising event during a war, but one cannot really know when it will occur and for how long. In the case of the Dutch famine, there is evidence that this event had an impact at the epigenetic level and that the development and subsequent life of people who were undernourished in the wombs of their mothers at this time was contingent *upon* that.

The central point of this chapter has been that our fate is not prescribed in our genes. Whereas genes are certainly important, there are so many phenomena taking place during development that may affect how the information encoded therein is used by cells. Genes do nothing on their own, but they are important resources for development. However, how these resources are read can be as important as the information encoded in them. Even people having the same genes can end up having different phenotypes, because these depend on which other genes are there (as in the case of eye color), on how tissues develop (as in the case of heterochromia), as well as on mutations and epigenetic changes that might occur during development. Overall, there is so much that can vary during development that you can take for granted that our fate, in terms of how we look and how we are (biologically speaking), is not prescribed in our genes.

Chapter 7

IT COULD BE HEREDITY, IT COULD JUST BE BAD LUCK

We often think of genetic diseases as also being hereditary ones; however, this is far from true. The first thing to note is that a disease can be neither inherited nor transmitted. In the case of infectious diseases, what is transmitted are infectious factors (viruses, bacteria, etc.), which affect the physiology of our body in various ways and destabilize it, thus producing the symptoms that we overall describe as the disease. Similarly, in the case of hereditary diseases what is inherited is not the disease itself but one or more DNA sequences that are somehow implicated in the development of the disease. But whereas hereditary diseases are often genetic for this reason, it is not the case that all genetic diseases are also hereditary ones. The reason for this is that the particular DNA sequences implicated in a disease may emerge anew in one's body due to mutations and not be inherited from one's parents. As I have already explained in the previous chapter, simple changes in DNA sequences, or larger ones that extend to parts of chromosomes, are possible. These genetic changes can take place not only during the production of reproductive cells but also during one's development. If we add to this the epigenetic changes that are also possible during one's development, then we can conclude that disease is contingent upon particular events.

Here is an interesting representative and provocative example of the contingencies of a well-known disease. In 1938, John Burdon Sanderson Haldane, a prominent figure in genetics of that time, performed a genetic analysis for the inheritance of hemophilia in the family of Queen Victoria. Haldane explained that an "ultra-microscopical particle called a gene" on the X chromosome was somehow involved in the production of a substance that was necessary for blood coagulation: in the case of women who have two chromosomes X, one inactive allele could be compensated by the action of the other active one; however, men have only one X chromosome, so any man with the inactive allele would be a hemophiliac.[1] Haldane then noted that one son, three grandsons, and six great-grandsons of Queen Victoria had been hemophiliacs, whereas her father, grandfather, and uncles all lived long enough to indicate that they did not have the disease. By analyzing their family tree, Haldane concluded: "The gene must have originated by mutation, and the most probable place and time where the mutation may have occurred was in the nucleus of a cell in one of the testicles of Edward, Duke of Kent, Victoria's father, in the year 1818."[2] This was a **turning point** for the royal families of Europe because several of their members suffered from the disease, to the extent that hemophilia was described as a "royal disease.". This mutation was contingent *per se*, as it was totally unpredictable. Furthermore, subsequent events were contingent *upon* that event, as several of Queen Victoria's descendants were hemophiliacs. As no known descendants of Queen Victoria who carry the disease have been studied, the type of hemophilia (A or B) has never been identified. However, the discovery and the identification of the presumed remains of the family of Czar Nicolas II have shown that it was type B hemophilia.[3]

In the so-called monogenic diseases, in which particular alleles of a single gene are considered responsible for the disease, it is definitely the case that people with the disease more often than not carry those particular alleles. These are therefore related to the disease, and hereafter I call them disease-related alleles. However, even in the case of typical monogenic diseases, the opposite conclusion does not always hold, as it is possible for one to carry the disease-related alleles but nevertheless not exhibit the disease. A classic example is β-thalassemia, which is often described as a recessive, monogenic disease (i.e., one needs to have received two disease-related alleles, one from each parent, in order to have the disease). Nevertheless, even in these cases it has been found that carrying the disease-related alleles does not entail that one will definitely have the disease, as it is important to also consider other genes. In the case of β-thalassemia, the problem is that one has a reduced amount of β-globin chains or no β-globin chains at all. This is due to mutations within or related to the expression of the *HBB* gene. The β-globin chains are part of a complex protein, hemoglobin, which transfers oxygen to tissues and transfers carbon dioxide away from them. When there is a reduced number of β-globin chains, then another constituent of hemoglobin, α-globin chains, are found in excessive numbers and aggregate in blood cells, causing either their abnormal development or their destruction. This, in turn, brings about the symptoms of β-thalassemia, which include a pale appearance and slowed growth and are described as anemia. However, people who have simultaneously developed a form of α-thalassemia (which is also due to mutations in genes *HBA1* and *HBA2* that limit the production of α-globin chains), or who have an increased production of hemoglobin F (normally produced in fetuses) in whom γ-globin chains bind some of the excessive α-globin chains, eventually have less α-globin chains in excess and therefore less severe anemia.[4]

Let us explore this phenomenon further. The gene that encodes the production of β-globin is found on human chromosome 11, whereas the genes that encode the production of α-globin are found on chromosome 16—let's consider one of them only. There are different kinds of molecular changes that may result in reduced production of α-globin and β-globin chains. Let us for simplicity call the alleles that produce "normal" amounts of these proteins α and β, and the mutated ones α(-) and β(-). It is possible that one's parents each carry one β(-) allele, and they thus have reduced amounts of β-chains that nevertheless do not significantly impact their lives. However, it is also possible that one of their children will come to have both of these alleles and thus develop a more severe form of β-thalassemia. However, it is possible that the mother also carries an α(-) allele, which also does not have a significant impact on her life, especially as she also has a β(-) allele. This woman has less α and β chains than normal, and a reduced amount of hemoglobin, but she faces no significant problems. Now whereas each one of the offspring of this couple each has a 25 percent probability of inheriting both β(-) alleles (see figure 7.1), the probability for each one of them to inherit the α(-) allele is 50 percent (see figure 7.2). These probabilities mean that if the couple had 100 offspring, 25 of them would have two β(-) alleles, whereas 50 of them would have an α(-) allele. These two events are independent, so in order to estimate the probability that one of the offspring would have two β(-) alleles and an α(-) allele, we have to multiply the above probabilities: 25% × 50% = 0.25 × 0.50 = 0.125 or 12.5%. This means that if this couple had 100 offspring, 12 or 13 among them would have two β(-) alleles and an α(-) allele.

We should be careful in thinking about probabilities here. This and any couple will not have 100 offspring, and we should keep in mind that women normally produce one ovum each month. What these probabilities practically indicate is that among those children who will inherit two β(-) alleles and who might thus develop a more severe form of β-thalassemia, half will also have the α(-) allele and might thus develop a milder form of the disease. In other words, about half of the offspring who would be expected to have β-thalassemia, would likely have mild or no symptoms because of the simultaneous presence of the α(-) allele. Whether or not this will be the case will depend on how the maternal chromosomes 11 and 16 will be combined during meiosis and whether the ovum produced will come to have both the β(-) and the α(-) alleles. Of course, we should not forget that the offspring might have none of these alleles, whereas some might have the same health condition with their parents (see figure 7.3). This is why, as already explained in chapter 5, the moment of the alignment of chromosomes during the meiotic division that produces the ovum is a **turning point** for the development of disease. Which combination of genes the offspring will receive is contingent *per se*, and the ensuing health condition will be contingent *upon* that event.

The possible outcomes in figure 7.3 clearly show that, more broadly, whenever a "monogenic" disease runs in a family, it is not possible to tell in advance whether the offspring will also have it. This is contingent *per se*, as it is not possible to know in advance which combinations of chromosomes and alleles one will have. Once the embryo is formed, whether or not it will develop the disease is contingent *upon* the chromosomes and alleles that it has. So the couple under discussion who carry the alleles and who have mild or no anemia, might have children who do not carry any disease-related alleles (e.g., O1), or children who are carriers like themselves and might have mild anemia (e.g., O2, O5, O7), or children who suffer severe anemia (e.g., O6). The important point to note is that finding that a child carries two disease-related β-alleles does not entail that it will develop the disease. This could be the case for child O6, but child O4 might have mild anemia because of the simultaneous presence of the α-allele. This also shows that there are no single genes "for" diseases, and that one would need to look at the alleles of different genes that may be associated with a disease. So the next time you worry about a "monogenic" disease that runs in the family, you should remember that it is not the disease but the disease-related alleles that are inherited, as well that even if both of your parents were carriers, this does not entail that you will have the disease-related alleles.

The typical research approach in the past has been the study of disease-related alleles in individuals with the disease. However, recent studies have shown that it is possible for individuals to have alleles directly related to a disease without exhibiting the symptoms usually associated with it. The available methods that make possible the sequencing of whole genomes have shown that people can have nonfunctional alleles without exhibiting the disease that one would expect to see, even for genes that have a well-established connection to a disease. It is also interesting that people who have both alleles of those genes in a nonfunctional state have different features than those people who have one functional allele. It has been estimated that each one of us may have about one hundred nonfunctional alleles, with about twenty genes being completely inactivated. In a study on the genomes of more than 100,000 people from Iceland, it was

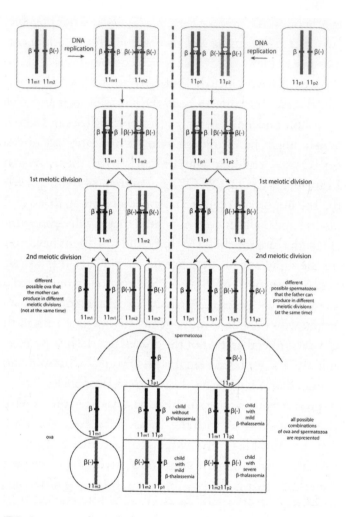

Figure 7.1. Two parents who are both carriers of the allele related to β-thalassemia can produce two types of ova and spermatozoa: those that have the disease-related allele and those that have the normal one. Theoretically, different kinds of offspring can emerge, and the couple can have either children like themselves, or children who do not carry the disease-related allele, or children who have two such alleles and are likely to develop the disease (it should be noted that the table does not reflect the probabilities before each fertilization and conception, as women only produce one ovum per cycle; it should also be noted that the second meiotic division in the ovum actually occurs after fertilization, but to avoid confusion here these are presented as independent events).

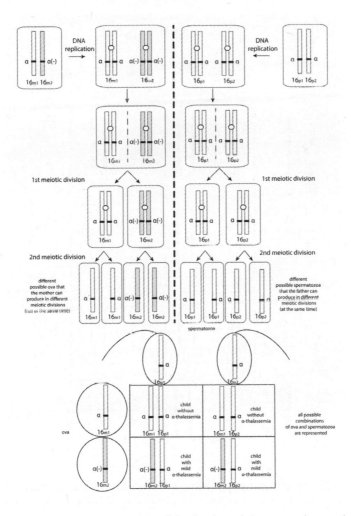

Figure 7.2. The mother is a carrier of the allele related to α-thalassemia and can produce two types of ova: those that have the disease-related allele and those that have the "normal" one. In contrast, all spermatozoa will have the normal allele. Theoretically, different kinds of offspring can emerge, and the couple can have children like themselves, either carrying the allele related to α-thalassemia, or children that have two normal alleles (it should be noted that the table does not reflect the probabilities before each fertilization and conception, as women only produce one ovum per cycle; it should also be noted that the second meiotic division in the ovum actually occurs after fertilization, but to avoid confusion here these are presented as independent events).

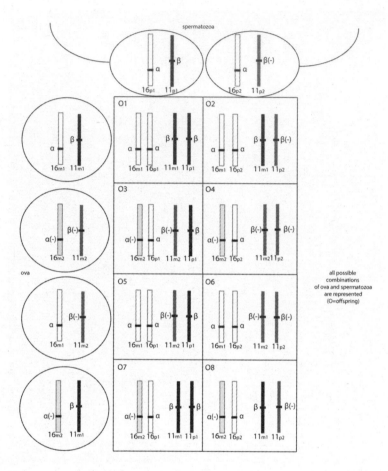

Figure 7.3. The father carries the allele related to β-thalassemia and can produce two types of spermatozoa: those that have the disease-related allele and those that have the "normal" one. The mother carries the allele related to β-thalassemia as well as the allele related to α-thalassemia. Therefore, theoretically, she can produce four kinds of ova: having either both of these alleles, or one of them, or none of them. As a result, several different kinds of offspring are possible: some will be completely healthy (e.g., O1); others will be carriers of either the α(-) or the β(-) allele (e.g., O2, O7); and some will likely have β-thalassemia (e.g., O6); but there will also be some who may not have significant problems, because they have both the α(-) and the β(-) allele (e.g., O8). (It should be noted that these table do not reflect the probabilities before each fertilization and conception as women only produce one ovum per cycle.)

concluded that about 8,000 individuals had one gene completely inactivated due to mutations.[5] It was also found that the occurrence of this phenomenon depended on the tissue, as completely inactivated genes were found less frequently in the brain and the placenta, and more frequently in the testis and the small intestine. An analysis of 874 genes, which were considered to cause 584 distinct severe monogenic disorders, in the genomes of 589,306 individuals identified thirteen adults who had mutations for eight severe monogenic diseases without exhibiting any symptoms. Apparently, the effects of these mutations were somehow compensated in these individuals.[6] All in all, this means that having two "bad" mutations is not by definition lethal, and the example of α- and β-thalassemias discussed above provides an explanation why.

Therefore, an important point to note is that most genetic diseases have multiple causes, that is, many different factors may cause them, and their occurrence depends a lot on critical events during our lives. This is why a genetic disease is not necessarily a hereditary one. An exemplar case of this are cancers. Cancers are diseases characterized by the development of tumors, overgrown masses of one's own cells, which can be solid in organs such as breasts, the colon, and lungs, or fluid in blood. Tumors are usually the outcome of alterations of the normal course of development and of the efficiency of the tissue-repairing processes. The problem that tumors cause is that their cells proliferate by consuming the resources of the body to the extent that the neighboring "normal" cells of the tissue die out. If tumors develop quickly and invade other tissues, they can cause the malfunction of organs and eventually death. Even though tumors can develop in a variety of organs and tissues, and researchers have identified more than one hundred different types, it seems that they have some common features, as particular physiological changes occur in all of them, although not necessarily in the same order. These changes occur in processes related to cell growth and death, and all together contribute to the transformation of healthy human cells into malignant ones, their uncontrolled proliferation and eventually the development of tumors in various tissues and organs. Normal human cells can become cancerous after many different kinds of changes occur, the latter being the outcome of successive mutations that happen in the same or neighboring cells.[7]

What is important is that the majority of cancers are caused by changes in our genome that in turn are caused by particular mutagenic factors to which we are exposed during our lives. These factors can be physical (e.g., α-, γ-, X- rays), chemical (e.g., tobacco smoke, DDT, asbestos), or biological (e.g., viruses). There are approximately 140 genes in which cancer-related mutations can occur. These genes relate to important cellular processes involved in the differentiation of cells, in the survival of cells, and in the maintenance of the genome. Overall, it seems that most human cancers are caused by two to eight sequential mutations that occur over the course of twenty to thirty years. Each of these mutations can confer a selective growth advantage to the cell in which it occurs, and so this cell may further proliferate against the normal cells. Any of these mutations is previously unpredictable. At the same time, all subsequent events depend on it, as a mutation that affects a cellular system that in turn results in abnormal cell proliferation disrupts cellular stability and makes the system more error-prone and thus more possible to mutate further. In this sense, each of these mutations is a **turning point**. Whether or not they will occur is contingent *per se*, and once they occur the subsequent outcomes are contingent *upon* them.

Figure 7.4 presents a concrete example from the well-studied colorectal cancer. A first mutation may occur in the *APC* gene, which has the information for the production of a protein that normally prevents cells from proliferating too fast in an uncontrolled manner. Such a mutation in *APC* may make the respective normal epithelial cell proliferate faster than the neighboring cells, thus producing a small adenoma (a benign tumor). This could happen between the ages of thirty and fifty. A second mutation may subsequently occur in the *KRAS* gene, which encodes the information for a protein that helps regulate the division of cells. Such a mutation may disrupt this regulation and further promote cell proliferation so that the small adenoma develops to a larger one. This second event may take place between, say, the ages of forty and sixty. Additional mutations to other genes such as *TP53*, which encodes a protein that normally regulates cell division by keeping cells from growing and dividing too fast or in an uncontrolled way, may eventually produce a malignant tumor that can also move to other organs. This could happen at any age between fifty to seventy. Whereas each of these mutations slightly increases the risk of cancer, consecutive mutations make the initial benign cells evolve to malignant ones that outgrow their neighboring normal cells, forming a malignant tumor (see figure 7.4). When its cells metastasize, for instance, to move to vital organs, the patient is likely to die.[8]

The mutations mentioned above that gradually transform normal cells to cancerous ones, and a normal tissue to a malignant tumor usually take place during cell divisions—in particular during the stage of DNA replication (refer back to figure 6.3). Therefore, except for the external factors that might cause these mutations, another important factor to consider is how often the cells in a tissue proliferate. To put it simply, the more the cells of a tissue divide, the more likely it is for mistakes during DNA replication, and thus mutations, to occur. Indeed, it has been estimated that the lifetime risk of several types of cancer is strongly correlated with the total number of cell divisions that normally take place in a tissue. This in turn explains why cancers develop in certain human tissues a lot more often than in others; if some tissues, like epithelia, divide more often than others, it is more likely for new mutations to occur. As a result, the variation in cancer risk, that is, the difference in having cancer in one or another tissue, is mostly due to variation in the number of cell divisions—specifically how frequently or how rarely these occur. For example, the same inherited mutant *APC* gene is responsible for the predisposition to cancers in both the large and the small intestines. However, it is about thirty times more likely for people to develop cancer in the large intestine than in the small intestine, and this could be due to the fact that there are about 150 times as many cell divisions in the former than in the latter. It also seems that only a third of this variation depends on differences in environmental factors or inherited factors.[9]

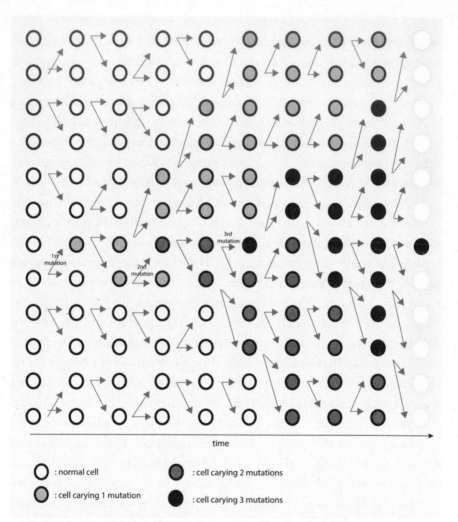

O : normal cell

O : cell carying 1 mutation

O : cell carrying 2 mutations

O : cell carrying 3 mutations

Figure 7.4. A depiction of the development of a tumor, which is actually an evolutionary process of natural selection among cancerous and normal cells. The former proliferate faster than the latter and are thus able to outcompete them. This becomes faster, the more mutations a cell accumulates. Thus, cancerous cells become prevalent in the tissue, which eventually is transformed to a tumor.

All of the above suggests that whereas mutations are certainly related to disease, there is not any kind of deterministic relation between the two. It is clear that particular mutations can bring about significant effects, and there is no question that even a single mutation can be enough to cause a disease. But whereas having a mutation can be a sufficient condition for having a disease, this does not mean that having a mutation automatically results in having a disease. It is important to recognize the complexities of gene expression and of the development of a disease, as well as the compensatory role that other genes can have. As shown in the case of thalassemias, people carrying alleles related to α-thalassemia and alleles related to β-thalassemia can have fewer clinical problems that those carrying alleles related to only one of them. Therefore, the combination of disease alleles that one receives from one's parents is critical. At the same time, a genetic disease is not necessarily an inherited one, as mutations also occur anew in people. Thus having, for instance, a family history of cancer can be indicative but not necessarily fatal. Unfortunately, for many cancers it is a question of what mutations occurred anew and when, and how our cells and tissues will deal with them. These new mutations, as well as the events that produce the various combinations of alleles that we all carry are genuine turning points.

Chapter 8

THOSE WHO DID NOT MAKE IT

Y ou have probably watched documentary films about how life
begins. How a spermatozoon (sperm cell) outcompetes the others
and arrives at the ovum (egg), which it then successfully fertilizes. After
that, the fertilized ovum travels down to the uterus, while its cells are
dividing and its morphology changes. Through a series of cell divisions,
the fertilized ovum becomes successively a two-cell, four-cell, eight-cell,
and sixteen-cell embryo, consisting of cells called blastomeres. Then the
blastomeres begin to form tight junctions with one another and further
divide, leading to a form called morula that consists of about thirty cells.
This further transforms to become a blastocyst, consisting of the inner
cell mass that will become the fetus and an outer layer, the trophoblast,
which will become the placenta (see figure 8.1). Eventually, the blastocyst
implants on the surface of the uterus. That is the beginning of a new life!

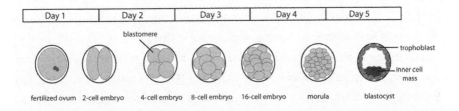

Figure 8.1. The development of the embryo from fertilization to implantation.

The cell divisions that take place during the development of the early
embryo, as well as after that, each result in cells that should be geneti-

cally identical to each other and to the cell from which they emerged. The process through which this happens is called mitosis or mitotic division (like meiosis, this is about the division of the cell nucleus, but I will refer to it as a cell division for the sake of simplicity). What happens during this process is simply that the DNA is replicated and then divided equally into two new cells. In this division, the homologous chromosomes, that is, the chromosomes that are very similar to each other in terms of their morphology and DNA sequence and that form pairs, do not meet up at the center of the cell but all line up and are divided into their chromatids. If you look closely at the second meiotic division (e.g., figure 5.2), it is exactly the same phenomenon that is taking place (and in this sense the mitotic and the second meiotic division are essentially the same process). As you can see in figure 8.1, during embryo development the total size of the embryo does not change and thus the cells composing the embryo become smaller and smaller. However, even though they may differ in size, the cells of the two-cell embryo and those of the blastocyst should be genetically identical, unless of course mutations have taken place (more about this below).

But what is often difficult to realize and even more difficult to estimate, is how many early embryos naturally fail to implant on the uterus and how many implanted embryos are naturally lost for various reasons. It is perfectly natural for an embryo to form in the mother's body but not implant on the uterus. This has no big consequences for the mother, certainly not physiologically, but perhaps some psychological consequences if she wants to have a child and it does not work out, but we need to keep in mind that it is possible. Unfortunately, it is also natural for embryos to implant but eventually fail to develop and be born. This process is referred to as a miscarriage, and it can have significant consequences for the mother, both physiological and psychological. This is life. Failure to implant or to develop and be born is a natural outcome in an early embryo's life, unfortunately.

Let us start with embryo loss during pregnancy, which is relatively easier to estimate. As mentioned above, the technical term for the

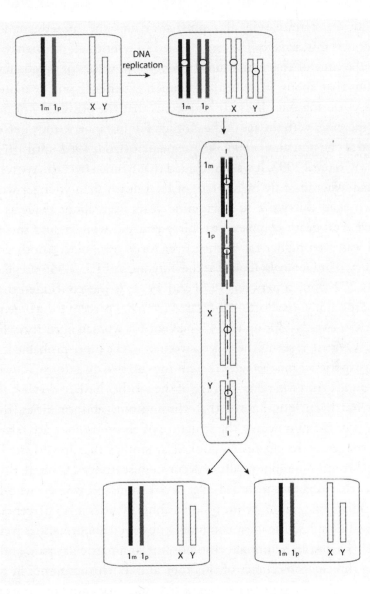

Figure 8.2. A depiction of the mitotic division, or mitosis, which occurs in the developing embryo and in general during cell proliferation. This process results in two new cells that should be genetically identical both to each other and to the cell from which they emerged, if no mutations take place.

spontaneous termination of a pregnancy and the subsequent loss of the fetus is miscarriage. The term is used to refer to all pregnancy losses from the time of conception until twenty-four weeks of pregnancy, and it seems that about 50 percent of women experience one or more miscarriages during their lives.[1] It also seems that the possibility of miscarriage increases with the age of the mother. For instance, a study in Canada looked at the outcomes of 94,346 pregnancies from 1961 until 1974 and from 1978 until 1993. It was concluded that women over thirty-five years old had a significantly higher rate of fetal death than younger women: women aged thirty-five to thirty-nine years were about twice as likely to have fetal death compared to thirty-year-old women, and this likelihood was even higher for women over forty years old.[2] Another study conducted in Denmark looked at the outcomes of 1,221,546 pregnancies of 634,272 women between 1978 and 1992. It was concluded that the risk of spontaneous abortion increased from 8.9 percent for women aged twenty to twenty-four up to 74.7 percent for women aged forty-five or older.[3] Overall, it seems the older a woman is, the more probable it is for her to experience miscarriage. These brings up two questions: Why is this happening? And what does the age of the mother have to do with that?

It has been long known that chromosome abnormalities, that is, changes in the numbers or the structure of chromosomes, are related to miscarriages. A recent review looked at studies that investigated how often chromosome abnormalities occur in miscarriages. Overall, thirteen studies that together included over seven thousand women who had a single miscarriage, found chromosome abnormalities in 45 percent of the analyzed samples. The most commonly observed abnormalities were trisomies (having an additional chromosome of a particular pair, and thus having three chromosomes of this type and 47 chromosomes in total), and they were found in 63 percent of the abnormal embryos. In six other studies, the findings were similar for over one thousand women who had experienced several miscarriages; compared to the 45 percent above, chromosome abnormalities were found in 39 percent of the samples analyzed. Once again, the most commonly observed abnormalities were trisomies, which were found in 65 percent of the abnormal embryos.[4]

When embryos do not have 46 chromosomes in their cells, but rather one more (trisomy) or one less (monosomy), these conditions are called aneuploidies, and they are generally considered to be due to mistakes in the first meiotic division of the ovum.[5] Normally, each ovum should have one chromosome of each pair and therefore 23 chromosomes in total (recall figure 5.3). However, when a mistake is made during the first meiotic division, for example, if there is a mistake in how the two homologous chromosomes 16 of a woman segregate, the outcome can be an ovum with two chromosomes 16 instead of one. This failure of chromosomes to segregate normally is called nondisjunction. Due to this, when fertilization occurs, the paternal chromosome 16 is added to the two maternal ones and the outcome is a trisomic embryo, that is, an embryo with three chromosomes 16 (see figure 8.3). A mistake can also occur during the second meiotic division if there is a nondisjunction of the two chromatids of the same chromosome (see figure 8.4). There are several possible ways how this can occur but for our discussion here it is enough to keep two in mind: (1) the nondisjunction of two homologous chromosomes in the first meiotic division, which results in cells with additional or missing whole chromosomes; and (2) the nondisjunction of the two chromatids of the same chromosome in the second meiotic division, which results in cells with additional or missing sister chromatids.[6] In most cases, aneuploidies result in miscarriages. However, there exist trisomies in which the embryo can be born and live for several years, albeit with various problems. Perhaps the most well-known case is trisomy 21, for which the respective condition is described as Down syndrome.

Miscarriages due to aneuploidies are relatively common in later stages of development, affecting about 5 percent of pregnancies. However, there is evidence that aneuploidies are even more prevalent at the beginning of pregnancy, that is, before implantation. Thus, researchers in the United Kingdom investigated the prevalence of aneuploidies in embryos before implantation, from fertilized ova to blastocysts, obtained from couples undergoing in vitro fertilization. In particular, the researchers studied 420 fertilized ova, 754 embryos at the stage of 8–16 cells (also called cleavage stage), and 1,046 blastocysts (see figure 8.1 for the main stages of human embryo development before implantation). From the 420 fertilized ova, only 108 were normal and had 23 pairs of chromosomes, with 312 of these having abnormal chromosome numbers. Approximately one third of all of these fertilized ova had three of more distinct aneuploidies. Most of these (97 percent) involved whole chromosomes, most frequently chromosomes 16, 21, 22, 15, and 19. Among the 754 cleavage-stage embryos, 130 had a normal chromosome number, 314 had one or two aneuploidies, and 310 had three or more aneuploidies. The most common abnormalities were in chromosomes 22, 16, 19, 21, and 13. Finally, among the 1,046 blastocysts studied, 438 were normal. The chromosomes that were most commonly found in aneuploidies in this case were chromosomes 22, 16, 15, 21, and 19.

The results of this study show the high frequency of chromosome abnormalities in pre-implantation embryos: 74 percent of the fertilized ova, 82 percent of the cleavage embryos, and 58 percent of the blastocysts studied. For all three groups, the observed abnormality rate was higher for older women (average age: forty-one years) than for younger women (average age: thirty-five to thirty-six years). An interesting finding of this study was that the fertilized ova exhibiting abnormalities due to mistakes in the second meiotic division were more than those exhibiting abnormalities due to mistakes in the first one. This is possible to discern, because a mistake in the first division results in an embryo with three different chromosomes of the same kind (see figure 8.3), whereas a mistake in the second division results in an embryo with two different chromosomes of the same kind, one of them existing twice (see figure 8.4). Abnormalities due to mistakes in the first meiotic division are found more often in younger women (younger than thirty-six), but in older women mistakes in the second meiotic division seem to take place more often than those in the first one.[7]

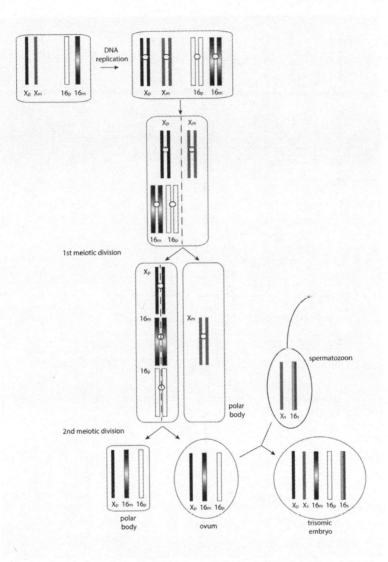

Figure 8.3. A mistake in the first meiotic division can result in an ovum with two chromosomes 16, and therefore a total of 24 chromosomes. When this is fertilized, a trisomic embryo—one with three chromosomes 16 and 47 chromosomes in total—will occur. It must be noted that the second meiotic division in the ovum actually occurs after fertilization, but to avoid confusion here these are presented as independent events.

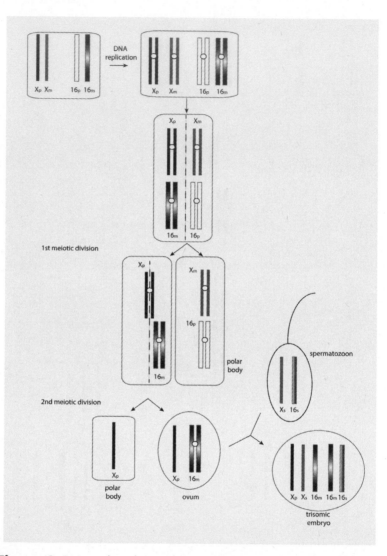

Figure 8.4. A mistake in the second meiotic division can result in an ovum with two chromosomes 16 of the same origin, and therefore a total of 24 chromosomes. When this is fertilized, a trisomic embryo—one with three chromosomes 16 and 47 chromosomes in total—will occur. It must be noted that the second meiotic division in the ovum actually occurs after fertilization, but to avoid confusion here these are presented as independent events.

But why do these mistakes seem to be related to the age of the mother? The meiotic divisions that result in spermatozoa start anew all the time and are completed within a few weeks; therefore, spermatozoa are normally newly produced cells. In contrast, the meiotic divisions that result in ova start (for all ova) before a woman is born, and they are completed only after ovulation (that is, the release of an ovum from one of the ovaries that occurs on average once a month around the middle of the menstrual cycle). Oocytes, the cells undergoing meiosis, enter a kind of arrest at the stage of the first meiotic division and may stay there for many years. In other words, the first meiotic division begins in the ovary of the woman when she is a fetus, and it is not completed until each ovulation, which occurs anywhere from ten to fifty years later. The consequence is that the older a woman is, the longer the oocytes have stayed in this arrest phase, and the more likely it becomes for a mistake in the segregation of the chromosomes of the same pair to occur. The exact mechanisms underlying this phenomenon are not entirely understood, but the connection is clear. Of course, there exist women who are forty years old (or older) and give birth to healthy babies. But these are rather the exception; rather than looking only at women of that age who are able to deliver babies, one should also look to all cases of miscarriage for women of that age.[8]

Another interesting question is why aneuploidies lead to miscarriage. This is not entirely clear; however, it has been suggested that a process of natural selection may be taking place that eventually prevents a chromosomally abnormal embryo to be implanted and the pregnancy to begin. When implantation takes place, the trophoblast cells of the blastocyst (see figure 8.1) start invading the surface of the uterus, called the endometrium. The cells therein have turned to specialized cells, called decidual cells, that are able to regulate the invasion of the trophoblast, to resist inflammatory and oxidative episodes, and to locally diminish the response of the immune system of the mother so that it won't destroy the embryo. The process of specialization of the cells of the endometrium (called decidualization) is thus critical in order for implantation to occur and for pregnancy to begin.[9] In humans, the decidualization of the endometrium is initiated toward the last part of the menstrual cycle, independent of the presence or absence of a pregnancy. This process is critical, as it provides the cells of the endometrium with the ability to recognize and eliminate abnormal embryos in the course of implantation. It seems that abnormal human embryos engage in intense signaling upon implantation, which in turn triggers a profound regional endometrial response. Decidualized endometrial cells are not passive, just waiting for the invasion of the blastocyst's trophoblast cells. Rather, in response to the signals of the blastocyst, they change and actually encapsulate the embryo. In doing so, they can identify developmentally impaired embryos and inhibit the production of molecules that have a key role in implantation.[10]

With all of the above, it should be clear that each of the two meiotic divisions is a **turning point** with regard to whether or not aneuploid embryos will emerge. The reason for this is that whether or not the mistakes in the disjunction of chromosomes or chromatids will occur, as well as exactly how they will occur, is unpredictable and thus contingent *per se*. One cannot know in advance whether the homologous chromosomes or the chromatids of the same chromosome will segregate during the first or the second meiotic division, respectively. Of course, researchers have concluded that this is more probable to happen to women older than forty. However, something more probable does not also become predictable. For every single mother-to-be over the age of forty, one cannot tell in advance what will happen, regardless of whether a trisomy is more probable to occur to her child than it is to that of a twenty-five-year-old woman. Neither can one know in advance to which of the resulting cells the nondisjuncted chromosomes or chromatids will go—to the ovum or to the polar bodies (see figures 8.3 and 8.4), and so one cannot predict the kind of mistake that will occur. Once a nondisjunction happens in any of the two meiotic divisions, the further development of the embryo is totally contingent *upon it*. An embryo with a trisomy at chromosome 21 could be viable, whereas one with a trisomy at chromosome 1 will not be.

So far, I have presented abnormalities occurring during the meiotic divisions producing the ovum, which were due to mistakes made before fertilization and before the zygote was formed. However, mistakes can also take place in the embryo itself, and it seems that they are actually quite common. These mistakes occur at the segregation of chromosomes during the mitotic divisions of the embryo, that is, the divisions that produce its various cells as it develops from the zygote to a two-cell embryo, to a four-cell embryo, to an eight-cell embryo, and so on. The outcome is described as chromosomal mosaicism, because the embryo consists of cells with different numbers of chromosomes, thus resembling a mosaic, even though these cells are derived from the same zygote. For instance, a study analyzed three-day-old and four-day-old human embryos to find the chromosome constitution of their blastomeres, that is, the cells these early embryos consist of. Of the twenty-three embryos, two were normal (having 46 chromosomes), and all of the others had different combinations of normal and aneuploid blastomeres, with some consisting exclusively of aneuploid blastomeres. About 91 percent of the human embryos studied were found to consist of a mixture of chromosomally normal and abnormal cells or of abnormal cells with different abnormalities. Whole chromosome aneuploidies were found in nineteen of the twenty-three embryos (83 percent). Of these embryos, three had the same chromosome aneuploidy in all of their blastomeres, which suggests that the mistake was made in the spermatozoon or in the ovum before the zygote was formed. All of the other aneuploidies had resulted from mistakes in the segregation of chromosomes during the cell divisions in the embryo.[11] Figure 8.5 shows an example of how this may happen. Imagine a two-cell embryo. If the division depicted in figure 8.2 takes place in one of the two blastomeres and the division depicted in figure 8.5 takes place in the other blastomere, then the outcome will be a four-cell embryo with cells that have different chromosome numbers. Two of the cells of that embryo will have 46 chromosomes (having emerged from the division in figure 8.2), but the other two will have 47 and 45 chromosomes, respectively (having emerged from the division in figure 8.5).

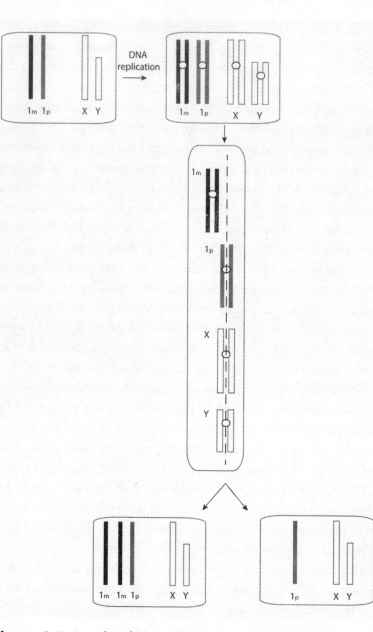

Figure 8.5. A mistake in the mitotic division may result in cells within the same embryo that have a different chromosome constitution.

A systematic review and meta-analysis of thirty-six studies of the chromosomal constitution of human pre-implantation embryos found that chromosomal mosaicism is quite prevalent. These studies were selected because, among other criteria, they provided the chromosomal constitution of each separate cell of the embryos analyzed. From a total of 815 spare embryos from IVF procedures, 177 (22 percent) were found to be diploid, 599 (73 percent) were found to be mosaic, and 39 (5 percent) were found to contain other chromosomal abnormalities. From the 599 mosaic embryos, 480 were found to be diploid-aneuploid mosaic (a mosaic embryo with one or more diploid, i.e., "normal," cells), and 119 to be aneuploid mosaic (a mosaic embryo without any diploid cells). From these results, one can conclude that mosaicism is by far the most common phenomenon in human pre-implantation embryos after IVF.[12]

Whether mosaicism is a phenomenon that occurs naturally or due to the IVF procedures is not clear. However, if it also occurs naturally, it would imply that whether or not an embryo will develop normally depends on the mitotic divisions therein. Therefore, each of the mitotic divisions that occur in the pre-implantation embryo is a **turning point** for its development. The reason for this is that whether or not these mistakes will occur, at which stage they will occur, as well as exactly how they will occur is unpredictable and thus contingent *per se*. One cannot know in advance whether the chromatids of the same chromosome will segregate during the mitotic division. Neither can one know at which stage of embryonic development (two-cell, four-cell, etc.) these mistakes will happen, as well as to which of the resulting blastomeres the nondisjuncted chromatids will go. Once a nondisjunction occurs in any of these divisions, the further development of the embryo will be totally contingent *upon* it. If a mistake like that in figure 8.5 occurs in the eight-cell stage, then one eighth of the cells of the embryo will have the chromosome abnormality. But if the mistake occurs at the two-cell stage, then half of the cells of the embryo will have the abnormality (see figure 8.6). Of course, if such a mistake does not occur at all, the embryo will be normal (figure 8.1).

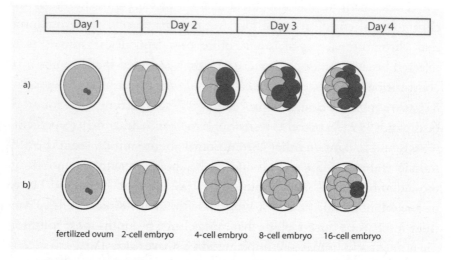

| Day 1 | Day 2 | Day 3 | Day 4 |

a)

b)

fertilized ovum   2-cell embryo   4-cell embryo   8-cell embryo   16-cell embryo

Figure 8.6. Mosaicism in human embryos; the earlier in development a mistake in the mitotic division occurs, the more blastomeres will be affected (affected cells are shown in a darker color here).

All of the above point to the conclusion that you and I are among the very few lucky ones who made it to birth. It seems that it is quite likely for ova and embryos to end up having abnormal chromosome numbers and, eventually, to fail to develop into fetuses. Because we do not have access to the actual events of conception, there is no way we can know what exactly is happening. But there is sufficient evidence for the conclusion that aneuploidies could be quite common in humans and therefore that there are many individuals who were conceived but did not make it to birth. In all cases, whether or not this will happen is unpredictable; and once it occurs, the outcome thereafter depends on the kind of mistake made. The older the mother (and the ovum under division) is, the higher is the probability for an aneuploid embryo to occur. But even in this case, one cannot predict whether an aneuploidy will actually occur, as is evident in the cases of women over the age of forty who give birth to healthy babies.

In this part of the present book, I have shown that whether or not we will be born, how we will look and be, and whether or not we will develop a genetic disease all depend upon particular turning points during our development. Neither any of these outcomes nor any developmental fate is in any way prescribed in our genes. Rather, all outcomes depend on critical events that are contingent *per se*, and *upon* which all subsequent outcomes are also contingent. As a secondary point, I cannot refrain from noting that even though many people have heard about mitosis and meiosis at school, it is likely that most of them have not realized how important and critical for the respective outcomes these processes are. Biology courses must relate the phenomena presented to real-life events and outcomes in order for people to make sense of them and understand their significance. This is what I have tried to do here. Considering all the phenomena presented in the four chapters of this part of the book, you have hopefully realized how contingent our presence in this world is (see figure 8.7).

Before moving on to part 2 to explore the kind of turning points occurring after we are born, a clarification might be useful. The turning points presented in chapters 5 through 8 relate to molecular/cellular events and phenomena. Whereas we can certainly identify the turning points in these phenomena, in many of these cases we cannot explain all of the details of these critical events. For instance, whereas we can say that the fact that a given spermatozoon fused with a given ovum during fertilization was a turning point for me to be born, we cannot tell why exactly it was the particular spermatozoon that made it to the ovum before the others, or why the particular ovum and not another matured one was released from the ovary. Similarly, whereas we can tell that a nondisjunction in the meiosis that resulted from the ovum was a turning point for a child with trisomy 21 to be born, we cannot tell why this nondisjunction took place at that particular moment. Nevertheless, we do not need to know all of their details in order to understand the significance of these turning points. Even if we cannot explain in perfect detail why the actual outcomes occurred and not any other among several possible outcomes, we can nevertheless recognize the significance of each actual event once it occurred, and therefore why it is a turning point (remember: the subsequent events were contingent *upon* that actual event that was itself contingent *per se*). The possibility of alternative events that could have brought about alternative outcomes (figure 8.7) is enough in order for us to recognize the importance of a turning point. How each one of us looks and is, as well as the presence of each one of us in this world, is the result of contingencies.

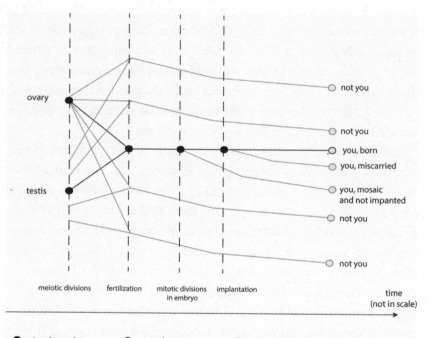

Figure 8.7. Turning points in human development. The birth of each one of us is the outcome of particular critical events during the production of the reproductive cells of our parents, fertilization, cell divisions in the embryo, and subsequently the embryo's implantation. Multiple possible routes begin from the testis because multiple spermatozoa are produced (but only four are shown here). Any of these could have fertilized the ovum and different offspring could have emerged accordingly, but they would not be you. However, the actual events that took place have led to your birth.

Part 3

TURNING POINTS IN HUMAN LIFE

Chapter 9

THE MOST IMPORTANT
EVENT IN LIFE

I t is often thought that facts are waiting in nature to be discovered by scientists, and that what makes the difference is who will discover them first. However, this is not accurate. Knowledge is in fact constructed, not discovered, and the contributions of particular individuals might make a vast difference both in the direction that the construction process will take and in the final outcome. Even though knowledge can be considered an impersonal resource that exists in books and webpages, and that can be acquired and used by anyone, it is not actually like this. This is the case for facts but not for knowledge; knowledge does not exist independently of a person who knows. Here is an example to illustrate the difference. Imagine that you have a sealed box that contains a coin. You may shake the box for a while and then put it on a table. The coin inside the box will land with either heads or tails up. Imagine that it lands heads up. This is a fact; it is not knowledge. Why? Because the box is sealed and you cannot actually see this. In other words, because you do not have access to this fact, you cannot know that the coin has landed heads side up. Knowing this requires some kind of access of your senses to this fact, which is impossible to have—unless you open the box. This is why knowledge always depends on the existence of a person who performs the act of knowing, turning a fact into knowledge.[1]

For this reason, scientific knowledge depends on the person who constructs it. It is one thing for someone to have access to the phenomena and facts of nature, and another to interpret them and turn them to knowledge. How these facts will be interpreted and what form of knowledge they will become depends on the conscious mind that will take over this task. Our knowledge of evolution still depends a lot on Charles Darwin's insights, concepts, and metaphors, despite the evolution of evolutionary science during the almost 160 years since he published his theory in his book *On the Origin of Species by Means of Natural Selection*.[2] In this chapter I explain why and how this theory was published in 1859 in this form, and neither in another time, nor in another format. A recent book under the title *Darwin Deleted* has been devoted to describe what might have happened in the history of evolutionary theory had Darwin never existed. Would we have arrived at the same theory, and, if so, when? Would someone else have developed it? One conclusion made in that book is that Darwin had a unique combination of interests that helped him develop the theory in the way he did. No one else could have done it in the same way and at the same time. As a result, a theory of evolution by natural selection could have only been developed by someone else much later.[3] Of course, simply saying that someone else, perhaps Alfred Russel Wallace, who is often described as the "co-discoverer" of natural selection, would have come up with a similar theory, is not enough. One must also answer questions such as: When? In what form? Including which arguments? Having what factual basis? With what impact? These are questions that are difficult to answer.

In the preface of one of his books, Wallace himself actually wrote:

I have felt all my life, and I still feel, the most sincere satisfaction that Mr. Darwin had been at work long before me, and that it was not left for me to attempt to write "The Origin of Species." I have long since measured my own strength, and know well that it would be quite unequal to that task. Far abler men than myself may confess, that they have not that untiring patience in accumulating, and that wonderful skill in using, large masses of facts of the most varied kind, that wide and accurate physiological knowledge, that acuteness in devising and skill in carrying out experiments, and that admirable style of composition, at once clear, persuasive and judicial, qualities, which in their harmonious combination mark out Mr. Darwin as the man, perhaps of all men now living, best fitted for the great work he has undertaken and accomplished.[4]

These words may make one think that the publication of *On the Origin of Species* was inevitable, that someone—perhaps Wallace if not Darwin— would write it anyway, and therefore that the only question was who that person would be. However, in the chapters in this third part of the book I argue that the foundations of modern evolutionary theory were set in 1859, and in the form of *On the Origin of Species*, because of particular turning points. In this sense, *On the Origin of Species* was not at all inevitable. A different theory could have been published years earlier or later in a different form, with significant differences in its content and readability and therefore its impact. The theory might as well have never been published, or might have never been conceived. Darwin was not *a priori* best fitted to do this in any special way. He indeed combined the available knowledge in a unique and original way, and arrived at well-established conclusions. But this happened only because Darwin took particular decisions, and because particular critical events oriented him in that direction.

Therefore, in the present book, I am not concerned with whether anyone else would have come up with the same theory, had Darwin never existed. Rather, I am interested in the different forms in which his theory could have been published and in the timing of these alternative publications. These are plausible scenarios to explore because Darwin wrote different manuscripts at different times. I therefore want to use the publication of Darwin's theory as a case study in order to point out why and how critical events that made things go in the direction they actually went, in order to show that they could have easily gone differently—whatever the alternative directions could have been. Whereas historians have argued that robust social processes generally influence outcomes and therefore individual events may not be sufficient to change the course of history,[5] in this chapter and those that follow I aim to show that Darwin's insights, choices, and decisions drove the process of scientific knowledge construction and, therefore, drove also the particular sequence of events to take place as it did and the particular outcome to be what it was. Thus, what I have in mind are the counterfactual scenarios of when and in what form Darwin could have published his theory. I identify some of the turning points toward the publication of *On the Origin of Species* based on my reading and interpretation of Darwin's autobiography. I do not claim to present neither the only, nor the most important, turning points of the path to *On the Origin of Species* by any external standards, but only based on Darwin's own recollections. In so doing I not only show that Darwin was not destined to write *On the Origin of Species* but also demonstrate the impact of turning points on the actual course of events.

Charles Robert Darwin was born in 1809 in Shrewsbury. His father was Robert Waring Darwin, a successful physician, investor, and son of the famous physician and poet Erasmus Darwin; his mother was Susannah Wedgwood, daughter of Josiah Wedgwood, founder of the famous pottery house and co-founder of the Lunar Society with Erasmus Darwin, Benjamin Franklin, James Watt, and others. Even though he enrolled in medical studies in 1825 in Edinburgh University, joining his brother Erasmus, he did not like medicine, as he detested the sight of blood and suffering. He preferred natural history and thus studied marine invertebrates with Robert Grant, a proponent of evolutionary ideas. Grant was an admirer of Jean-Baptiste Lamarck and his evolutionary views, and he introduced Darwin to them. However, even though Darwin "listened in silent astonishment," as far as he could judge when he wrote his autobiography, he did this "without any effect" on his mind.[6] By this he probably meant that he did not become a proponent of evolution at that time, despite his exposure to such ideas. Although he came from an open-minded, free-thinking family background, Darwin was not an evolutionist right from the start. He was aware of the writings of his paternal grandfather Erasmus, but as he died in 1802 and Charles was born in 1809, the two never met and there was no direct influence from grandfather to grandson. But Charles was especially interested in natural history from a very young age, and this was an important motivation for him to study nature.

Given his interests, his father suggested to Darwin that he study to become a clergyman, as an alternative to medicine. Darwin considered this an opportunity to continue studying natural history and was thus admitted as a student at Cambridge University in 1828. During his years there, Darwin became a passionate amateur naturalist. This passion is evidenced in his autobiography:

> But no pursuit at Cambridge was followed with nearly so much eagerness or gave me so much pleasure as collecting beetles. It was the mere passion for collecting, for I did not dissect them and rarely compared their external characters with published descriptions, but got them named anyhow. I will give a proof of my zeal: one day, on tearing off some old bark, I saw two rare beetles and seized one in each hand; then I saw a third and new kind, which I could not bear to lose, so that I popped the one which I held in my right hand into my mouth. Alas it ejected some intensely acrid fluid, which burnt my tongue so that I was forced to spit the beetle out, which was lost, as well as the third one.[7]

Darwin also became a close discussant with botany Professor John Stevens Henslow, who had a major influence on him:

> I have not as yet mentioned a circumstance which influenced my whole career more than any other. This was my friendship with Prof. Henslow. Before coming up to Cambridge, I had heard of him from my brother as a man who knew every branch of science, and I was accordingly prepared to reverence him. He kept open house once every week, where all undergraduates and several older members of the University, who were attached to science, used to meet in the evening. I soon got, through Fox [Darwin's second cousin, also a student at Cambridge], an invitation, and went there regularly. Before long I became well acquainted with Henslow, and during the latter half of my time at Cambridge took

long walks with him on most days; so that I was called by some of the dons "the man who walks with Henslow"; and in the evening I was very often asked to join his family dinner. His knowledge was great in botany, entomology, chemistry, mineralogy, and geology. His strongest taste was to draw conclusions from long-continued minute observations. His judgment was excellent, and his whole mind well-balanced; but I do not suppose that anyone would say that he possessed much original genius.[8]

This brings us to an important event in Darwin's life. It was Henslow who told Darwin that Commander Robert FitzRoy was looking for someone to travel aboard the HMS *Beagle*. This seemed to be a fascinating opportunity for Darwin, as he had already started thinking about the possibility of making such a voyage, visiting exotic places, and making at least a humble contribution to natural science:

> During my last year at Cambridge I read with care and profound interest Humboldt's *Personal Narrative*. This work and Sir J. Herschel's *Introduction to the Study of Natural Philosophy* stirred up in me a burning zeal to add even the most humble contribution to the noble structure of Natural Science. No one or a dozen other books influenced me nearly so much as these two. I copied out from Humboldt long passages about Teneriffe, and read them aloud on one of the above-mentioned excursions, to (I think) Henslow, Ramsay and Dawes; for on a previous occasion I had talked about the glories of Teneriffe, and some of the party declared they would endeavour to go there; but I think that they were only half in earnest. I was, however, quite in earnest, and got an introduction to a merchant in London to enquire about ships; but the scheme was of course knocked on the head by the voyage of the *Beagle*.[9]

In the meantime, Henslow also persuaded Darwin that he had to study geology. An important figure among his professors at Cambridge was Adam Sedgwick, professor of geology, who was soon planning to leave for a geological exhibition in North Wales. Henslow asked Sedgwick to accept Darwin to accompany him, and so he did. This was another important experience for Darwin, who wrote in his autobiography that "this tour was of decided use in teaching me a little how to make out the geology of a country. . . . On this tour I had a striking instance how easy it is to overlook phenomena, however conspicuous, before they have been observed by anyone."[10]

As soon as Darwin returned from his geological tour with Sedgwick, he received the news about the *Beagle* voyage. A letter from Henslow informed him that FitzRoy was willing to accommodate any young man who would volunteer to join the *Beagle* voyage as an unpaid naturalist. Darwin was eager to accept but his father strongly objected, and so Darwin wrote and refused the offer. But his father had also stated in his response: "If you can find any man of common sense, who advises you to go, I will give my consent." As it happened, on the next day Darwin got together with his uncle, Josiah Wedgwood, who was a person his father considered sensible and who eventually offered to go with Darwin to Shrewsbury to talk to his father. Josiah thought that it would be wise for Darwin to accept Fitz-Roy's offer. Thus a deal was sealed. Darwin then run to meet Henslow and FitzRoy in order to ensure that he would join the voyage.[11]

The *Beagle* voyage lasted for almost five years, from December 27, 1831, to October 2, 1836. During that time, the ship traveled around the world. Darwin was able to study geology and zoology in several places in South America, the Galápagos Islands and other islands of the Pacific Ocean. He was also able to collect specimens and send them back to England with other ships that he encountered at various places. His geological studies were extremely diligent, and when he returned to England, he was already considered an important geologist. During the *Beagle* voyage Darwin read the first volume of Charles Lyell's *Principles of Geology*. This book was extremely useful to him in various ways; he also had the opportunity very early during the voyage to experience phenomena that were in agreement with Lyell's conclusions. Lyell had suggested that the natural forces currently operating on Earth have always been at work, and the current status of the Earth is the outcome of slow, gradual, natural processes of change. Thus, erosion, earthquakes, and volcanos were sufficient to slowly bring about changes, without any need for miraculous events that would cause them.[12]

Darwin made significant observations during the *Beagle* voyage. Perhaps the most well-known are those concerning the various types of finches on the Galápagos Islands. Darwin visited the islands in September and October 1835, and collected various finches' specimens without really wondering about how they might be related to one another. However, after his return to England, he became interested in whether the finches he had collected were distinct varieties or distinct species. Darwin thought that in the case that these were distinct species, they could have simply migrated to the Galápagos Islands from the nearby American continent. But if they were distinct varieties, this could mean that a process of speciation could be taking place, as varieties have the potential to give rise to new species. An answer to this question was given in March 1837 when ornithologist John Gould told Darwin that the birds he had collected at the Galápagos Islands were distinct species, as well as that these species were not found on the American continent. Therefore, the only sound conclusion was that these species had originated on the Galápagos Islands. This was an astonishing information for Darwin, as the species that had originated on the volcanic Galápagos Islands were not similar to those found in other volcanic islands, that is in other places with similar environmental conditions, but were rather similar to species living under very different conditions to the nearby American continent. Therefore, the characteristics of those species and the similarities among them should mostly be due to their common ancestry, rather than to their adaptation to similar environments (exactly because the environment in the Galápagos Islands was very different from that of the American continent).[13]

This and other observations, such as the presence of several endemic species—that is, of species that were only found in that particular place and nowhere else in the world—on the Galápagos Islands, made Darwin undergo a significant conceptual shift from his original views. Initially, he had admired the arguments of William Paley (already mentioned in chapter 4), who had suggested that if organisms exhibited more complex features than human-made artifacts, the only explanation could be that they were made by an artificer who was even more competent than humans—and this could only be God. In his autobiography Darwin wrote:

> In order to pass the B.A. examination, it was, also, necessary to get up Paley's Evidences of Christianity, and his Moral Philosophy. This was done in a thorough manner, and I am convinced that I could have written out the whole of the Evidences with perfect correctness, but not of course in the clear language of Paley. The logic of this book and as I may add of his Natural Theology gave me as much delight as did Euclid. The careful study of these works, without attempting to learn any part by rote, was the only part of the Academical Course which, as I then felt and as I still believe, was of the least use to me in the education of my mind. I did not at that time trouble myself about Paley's premises; and taking these on trust I was charmed and convinced by the long line of argumentation.[14]

But after his experiences during the *Beagle* voyage, Darwin started to change his mind.

> The old argument of design in nature, as given by Paley, which formerly seemed to me so conclusive, fails, now that the law of natural selection has been discovered. We can no longer argue that, for instance, the beautiful hinge of a bivalve shell must have been made by an intelligent being, like the hinge of a door by man. There seems to be no more design in the variability of organic beings and in the action of natural selection, than in the course which the wind blows. Everything in nature is the result of fixed laws.[15]

However, it took Darwin some time to arrive at this view and to conceive of a natural process that could bring about adaptations. He did not just have a eureka moment during the *Beagle* voyage, but during it he made several crucial observations that set the foundations for the changes in his views.

But this is not the only, and perhaps not even the main, reason that the *Beagle* voyage was crucial for the development of Darwin's theory. It is also the habits of mind and work that he acquired during this voyage, which were very important because they guided his further work for the rest of his life. Thus, all subsequent events were contingent *upon* the *Beagle* voyage, because all his future work was shaped by this experience. And making the voyage itself was contingent *per se*. Henslow might not have made the suggestion to him, his father might not have been convinced by his uncle to consent to him traveling, or FitzRoy might have not made the offer at all. Darwin mentioned in his autobiography that he later heard that he was very close to being rejected by FitzRoy because of the shape of his nose that indicated that he did not have the energy and determination necessary for the voyage.[16] In this sense, it was a series of contingencies that made possible for Darwin to acquire this experience, which shaped his future work in science. This is why the *Beagle* voyage, which was contingent *per se* and *upon* which Darwin's future work was contingent, is a **turning point**. As Darwin wrote in his autobiography:

> The voyage of the *Beagle* has been by far the most important event in my life and has determined my whole career; yet it depended on so small a circumstance as my uncle offering to drive me 30 miles to Shrewsbury, which few uncles would have done, and on such a trifle as the shape of my nose. I have always felt that I owe to the voyage the first real training or education of my mind. I was led to attend closely to several branches of natural history, and thus my powers of observation were improved, though they were already fairly developed.[17]

> The above various special studies were, however, of no importance compared with the habit of energetic industry and of concentrated attention to whatever I was engaged in, which I then acquired. Every-

thing about which I thought or read was made to bear directly on what I had seen and was likely to see; and this habit of mind was continued during the five years of the voyage. I feel sure that it was this training which has enabled me to do whatever I have done in science. . . . That my mind became developed through my pursuits during the voyage, is rendered probable by a remark made by my father, who was the most acute observer whom I ever saw, of a sceptical disposition, and far from being a believer in phrenology; for on first seeing me after the voyage, he turned round to my sisters and exclaimed, "Why, the shape of his head is quite altered."[18]

The counterfactual question that naturally comes to mind is this: Would Darwin have developed the theory of natural selection had he not traveled aboard the HMS *Beagle*? One cannot really know; however, given Darwin's remarks above, it is clear that this experience shaped (besides his head as his father remarked) his thinking and thereafter the way that his theory was developed. It seems very likely that his views would have changed less radically or much later than how they did if he had not had the experiences during this voyage. But what matters more are the skills and habits of work that he developed. If he had lived in London, or anywhere else during those five years, there is no conceivable way that he would have devoted so much time on observations, note-taking, and interpretations. The limited social life aboard the ship, and the lack of other interactions with family or of a romantic relationship certainly gave him plenty of time to work and develop the skills and habits that guided all of his further work. Therefore, no matter what could have happened had he not traveled aboard the *Beagle*, it is important to note the importance of the *Beagle* experience for his subsequent work, explicitly acknowledged by Darwin himself.

His experiences during the voyage also left a lasting impression on him, both because of things he saw or experienced for the first time in his life and because he was thus able to make contributions to natural science.

The glories of the vegetation of the Tropics rise before my mind at the present time more vividly than anything else. Though the sense of sublimity, which the great deserts of Patagonia and the forest-clad mountains of Tierra del Fuego excited in me, has left an indelible impression on my mind. The sight of a naked savage in his native land is an event which can never be forgotten. Many of my excursions on horseback through wild countries, or in the boats, some of which lasted several weeks, were deeply interesting; their discomfort and some degree of danger were at that time hardly a drawback and none at all afterwards. I also reflect with high satisfaction on some of my scientific work, such as solving the problem of coral-islands, and making out the geological structure of certain islands, for instance, St. Helena. Nor must I pass over the discovery of the singular relations of the animals and plants inhabiting the several islands of the Galapagos archipelago, and of all of them to the inhabitants of South America. As far as I can judge of myself I worked to the utmost during the voyage from the mere pleasure of investigation, and from my strong desire to add a few facts to the great mass of facts in natural science.[19]

Therefore, it was both the pleasure of investigation and his ambition to be listed among the great men of science that fueled Darwin's persistence for doing scientific work. One cannot know whether Darwin would have worked with the same persistence and the same passion had he stayed forever in his homeland. Even if Darwin's actual persistence and passion was not entirely due to his *Beagle* experience, it was certainly fueled by it. As soon as Darwin realized the implications of his findings during the *Beagle* voyage, he started taking notes of his thoughts about transmutation—the modification of species (the term *evolution* was not used at the time in this sense). In his personal diary for 1837, he was explicit about how influential his conclusions from the findings during the *Beagle* voyage were:

In July opened first note book on "Transmutation of Species"—had been greatly struck from about month of previous March on character of S. American fossils, and species on Galapagos Archipelago. These facts origin (especially latter) of all my views.[20]

The impact of the *Beagle* voyage is also evident in the first phrases of *On the Origin of Species*:

> WHEN on board H.M.S. "Beagle," as naturalist, I was much struck with certain facts in the distribution of the inhabitants of South America, and in the geological relations of the present to the past inhabitants of that continent. These facts seemed to me to throw some light on the origin of species—that mystery of mysteries, as it has been called by one of our greatest philosophers.[21]

Therefore, a critical event toward the development of Darwin's theory was his five-year-long experience aboard the *Beagle*. His observations and the habits of mind and work he acquired during that time were critical for this subsequent work and the development of his theory, as Darwin himself acknowledged in his notebook right after the trip, in *On the Origin of Species* and in his autobiography. The voyage aboard the *Beagle* was an unpredictable experience with a long-lasting impact; it was contingent *per* se and Darwin's subsequent work was contingent *upon* it. In other words, it was a genuine turning point.

Chapter 10

A THEORY BY WHICH TO WORK

Afer the *Beagle* experience Charles Darwin had become convinced that species were not fixed and that transmutation was a fact of nature. Already in March 1837 he made relevant notes in the Red Notebook, about the possibility of a species changing into another, after considering the conclusions of ornithologist John Gould, already mentioned in the previous chapter. In July 1837, Darwin began Notebook B, in which he started taking notes of his thoughts and reflections on transmutation. Darwin used the title *Zoonomia* in that notebook, which literally means "the laws of life." This was the title of a book that his grandfather Erasmus had written and that contains evolutionary speculations (the subtitle of the book was *The Laws of Organic Life*).[1] Charles was not initially influenced by his grandfather's writings, in part because of their speculative character:

> I had previously read the *Zoönomia* of my grandfather, in which similar views are maintained, but without producing any effect on me. Nevertheless it is probable that the hearing rather early in life such views maintained and praised may have favoured my upholding them under a different form in my *Origin of Species*. At this time I admired greatly the *Zoönomia*; but on reading it a second time after an interval of ten or fifteen years, I was much disappointed, the proportion of speculation being so large to the facts given.[2]

Darwin wrote in his Notebook B his first reflections about transmutation.[3] The notebook includes reflections and answers to many questions about adaptation:

Changes not result of will of animal, but law of adaptation as much as acid and alkali.

Organized beings represent a tree <u>irregularly branched</u> some branches far more branched—Hence Genera. —)[4]

This requires principle that the permanent varieties produced by inter confined breeding & changing circumstances are continued & produced according to the adaptation of such circumstances, & therefore that death of species is a consequence (contrary to what would appear from America) of non-adaptation of circumstances.[5]

The condition of every animal is partly due to direct adaptation & partly to hereditary taint; hence the resemblances & differences for instance of finches of Europe & America. &c., &c., &c.[6]

Astronomers might formerly have said that God ordered each planet to move in its particular destiny. —In same manner God orders each animal created with certain form in certain country, but how much more simple & sublime powers let attraction act according to certain law such are inevitable consequen[ces].

Let animals be created, then by the fixed laws of generation, such will be their successors. —

Let the powers of transportal be such, & so will be the forms of one country to another. —Let geological changes go at such a rate, so will be the number & distribution of the species!![7]

In these excerpts, we read Darwin's view that adaptation is the result of a law of nature; that it is due to changing circumstances so that species either adapt or die out; and that therefore the features of every animal are due to either adaptation or heredity. Most important, he thought that natural processes are sufficient to bring about changes in species and divine intervention was not necessary, even though it might be God who had established these processes. These show that many of the elements of Darwin's theory were in place as early as 1837. But the important question was still unanswered: What was the natural principle behind these processes?

> With belief of change transmutation & geographical grouping we are led to endeavour to discover causes of change, the manner of adaptation (wish of parents??) instinct & structure become full of speculation & line of observation. . . . this & direct examination of direct passages of species structures in species, might lead to laws of change, which would then be main object of study, to guide our past speculations with respect to past and future. The grand question which every naturalist ought to have before him when dissecting a whale or classifying a mite, a fungus, or an infusorian is, "What are the Laws of Life."[8]

In Notebook C,[9] which Darwin started writing in February 1838, we also read his thoughts about domestication and artificial selection, and its importance for understanding the variation of organisms in nature. Darwin had read the pamphlets written by animal breeders, such as John Sebright and John Wilkinson, who were explicit about the nature and power of artificial selection:

> Sir J. Sebright—pamphlet most important showing effects of peculiarities being long in blood.++ thinks difficulty in crossing race—bad effects of incestuous intercourse. —excellent observations of sickly offspring being cut off so that not propagated by nature. —Whole art of making varieties may be inferred from facts stated. —++ Fully supported by Mr Wilkinson, = milking hereditary, developement of important organ (see mark on pages),—crosses of diff: breeds succeed. yet seems to grant that difficult & other go back to either parent. — Shows instinct (Sir J. Sebright admirable essay) hereditary journey wild ducks. —lose as well as gain instincts. Wild & tame rabbit good instance—instincts of many kinds in dogs as clearly applicable to formation of instincts in wild animals many species in one genus external circumstances in both cases effect it. —Sir J. Sebright excellent authority because written on dog. Barking—applies it to national character.[10]

Darwin thus thought that sustained selection for small changes could be taking place in nature, in a similar way with the way that breeders achieved this through the process of artificial selection. If breeders were able to artificially select and cross particular individuals and thus obtain offspring with the characteristics they wished, then it might be possible for a similar kind of selection to naturally occur in nature. It seems that the Sebright pamphlet stimulated Darwin to search for a mechanism in nature equivalent to the artificial selection employed by breeders (interestingly, in his pamphlet Sebright was quite explicit about natural selection and the analogy with artificial selection). Darwin also joined several pigeon-breeding clubs to see for himself how selective breeding could produce new varieties. It is important to note that Darwin was not the first to consider animal and plant breeding as a source for understanding natural history; furthermore, information on plant and animal breeding was widely disseminated in England at that time. However, he was the first to pay special attention to it and to fully develop the analogy that paved the way for the theory of natural selection.[11]

But an important element was missing. Artificial and natural selection certainly have similarities, and it was these similarities that made Darwin see the analogy between them. In both cases, there can be selection of particular organisms who reproduce whereas others do not. Thus, the characteristics of a population or species might gradually change exactly because of the selection that takes place, due to which only the characteristics of those that survive and reproduce are passed on to the next generation. However, there is also a very important difference. Artificial selection takes place when a conscious agent allows some organisms but not others to produce offspring, according to his or her own intentions. This was what breeders like Sebright did. However, no such conscious agent exists in nature. Therefore, a big problem for Darwin was to figure out how an unmediated, unintentional natural process might result in the selection of variants in the same way as with artificial selection. Because there could be no conscious intelligent agent that would miraculously intervene and make the selection, Darwin had to find a natural process that could have such a strong effect and that could bring about similar outcomes: the selective reproduction of particular individuals but not of others, so that the next generations would have some "selected" characteristics—that is, either characteristics intentionally selected by the breeders in the case of artificial selection, or characteristics that had provided their bearers with some kind of advantage in the case of natural selection.

Darwin found a solution to this problem in September 1838, when he read Thomas Malthus's *Essay on the Principle of Population*. In short, Malthus argued that while the tendency of human populations was to increase at a geometric rate $(2 \times 2 \times 2 \times 2 = 16)$, agricultural production increased at an arithmetic rate $(2 + 2 + 2 + 2 = 8)$. Consequently, there would inevitably be a struggle for resources that would limit the increase of human population because the resources would not be enough to support all members of the population. Darwin thought that a similar process could be taking place in nature. On September 28, he thus wrote in his Notebook D:

> We ought to be far from wondering of changes in numbers of species, from small changes in nature of locality. Even the energetic language of ~~Malthus~~ Decandolle does not convey the warring of the species as inference from Malthus. —increase of brutes must be prevented solely by positive checks, excepting that famine may stop desire. —in nature production does not increase, whilst no check prevail, but the positive check of famine & consequently death. Population is increase at geometrical ratio in far shorter time than 25 years—yet until the one sentence of Malthus no one clearly perceived the great check amongst men. —there is spring, like food used for other purposes as wheat for making brandy. —Even <u>a few</u> years plenty, makes population in Men increase & an <u>ordinary</u> crop causes a dearth. take Europe on an average every species must have same number killed year with year by hawks, by cold &c. —even one species of hawk decreasing in number must affect instantaneously all the rest. —The final cause of all this wedging, must be to sort out proper structure, & adapt it to changes. —to do that

for form, which Malthus shows is the final effect (by means however of volition) of this populousness on the energy of man. One may say there is a force like a hundred thousand wedges trying force into every kind of adapted structure into the gaps of in the oeconomy of nature, or rather forming gaps by thrusting out weaker ones.[12]

In his autobiography, Darwin recounted his reading of Malthus and how crucial it was for the development of his theory after he had realized the importance of selection:

> I soon perceived that selection was the keystone of man's success in making useful races of animals and plants. But how selection could be applied to organisms living in a state of nature remained for some time a mystery to me.
>
> In October 1838, that is, fifteen months after I had begun my systematic enquiry, I happened to read for amusement Malthus on *Population*, and being well prepared to appreciate the struggle for existence which everywhere goes on from long-continued observation of the habits of animals and plants, it at once struck me that under these circumstances favourable variations would tend to be preserved, and unfavourable ones to be destroyed. The result of this would be the formation of new species. Here, then, I had at last got a theory by which to work.[13]

Darwin actually did more than simply find a theory by which to work. First, whereas Malthus described the struggle of a whole species against its environment, Darwin also realized that there was another kind of struggle that resulted from the competition between individuals of the same species. Therefore, Darwin came up with the idea that it was the external environment, with the different types of struggle that it entails, that assumed the role of the conscious agent of artificial selection; this was the key concept that made natural selection strongly analogous to artificial selection. Second, Darwin expanded Malthus's concept of population checks from just the limitation of resources to include any factor of the environment that might limit population increase. In fact, Darwin made the idea of struggle the driving force behind adaptive change.[14]

Here is a simple illustration of this process. Let's imagine a population consisting of both gray and white beetles living in a "gray" environment (the environment cannot of course be gray, but let's really use our imagination). Imagine also that there exist birds in this area that feed on these beetles. As a result of this bird predation, there is a struggle for existence among these beetles. Now, as it happens, the gray beetles can conceal themselves in that particular environment better than the white ones can, because there is a contrast between the color of the latter and the color of the environment. Therefore, the white beetles are more visible to birds and so are more prone to be eaten by them. In contrast, the coloration of the gray beetles provides them with a relative advantage, as it is more difficult for the birds to spot and eat them. It is because of these differences that more gray than white beetles can survive and reproduce; and, therefore, there will be more gray and fewer white ones in the next generation. If this process continues for several generations, the white beetles might cease to exist and the initial population might evolve to a new population of gray beetles. The whole process by which some individuals but not others can survive and reproduce in the struggle for existence is natural selection. This is a natural, unconscious process of selection due to environmental conditions that are more favorable for the survival and reproduction of some individuals and less favorable for that of others (see figure 10.1). It must be noted that what evolves is the whole *population*, not the individual beetles themselves. A common misconception is that individual organisms change in order to adapt; in this case, one might think that the white beetles would change to become gray so that they would be better adapted, and thus the population would overall change. But this is not the case. A population adapts because some members survive and reproduce whereas others die out before reproducing and thus the characteristics of the former become more and more prevalent across generations, whereas those of the latter become less and less frequent.

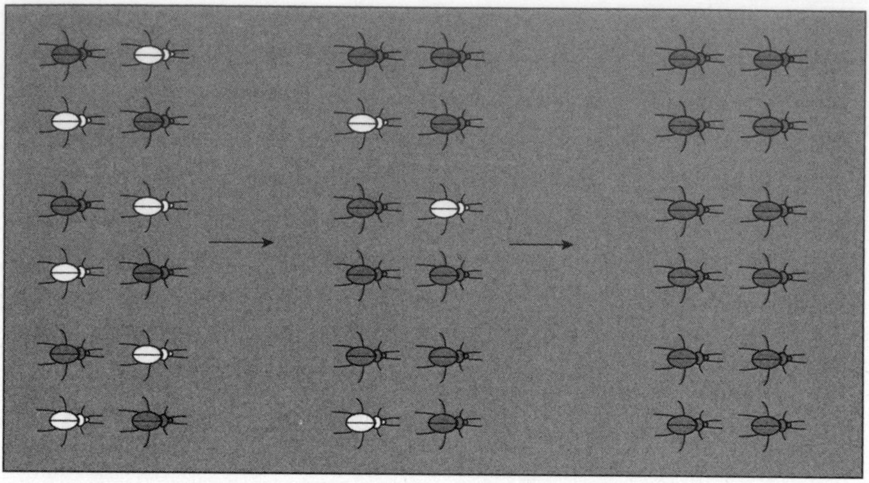

Figure 10.1. If certain individuals in a population have an advantage in survival and reproduction (in this case, the gray beetles are better concealed than the white ones and therefore are less visible to bird predators), the population may eventually evolve so that it consists of such individuals only because, in the struggle for existence, the others will die out before reproducing. This process is called natural selection.

Why were Malthus's ideas so crucial for Darwin's theory? The answer is that Malthus's concepts provided Darwin with quantitative laws that were in accordance with the predominant philosophy of science of the time. Two major figures of this philosophy were John Herschel and William Whewell. They both noted that good science required an extensive evidential basis and the identification of a mechanism or cause that could explain phenomena in different areas. In particular, they believed that the aim of science was to find the laws of nature and then to identify the true causes that guided the workings of these laws. Darwin was eager to work according to the philosophy of science and the scientific standards of his time. He knew Whewell from his years at Cambridge and also met Herschel in South Africa during his *Beagle* voyage. So, as stated above, the work of Malthus provided Darwin with quantitative laws, a requirement according to Herschel and Whewell, on which the idea of the struggle for existence could be based, and, in turn, could serve as the basis for the process of natural selection. One of Darwin's main concerns was to establish natural selection as a true cause. Therefore, he used the analogy with artificial selection in order the establish the existence and competence of natural selection to produce new species. However, he never managed to show, with observations nor with experiments, that natural selection was actually competent to produce new species.[15]

I must note at this point that this did not happen all at once in 1838. Darwin's reading of Malthus was crucial for his conception of the idea of natural selection, but the full theory was not developed in one fell swoop. Overall, it seems that there are two important shifts in Darwin's views: (1) the shift from special creation to transmutation and then to natural selection, which was completed soon after Darwin read Malthus; and (2) the shift from perfect adaptation to relative adaptation, which took many more years to be completed (more about this in the next chapter). One might assume that the first shift, from special creation to natural selection, was the most important one. After all, this is when Darwin came up with the fundamental idea of his evolutionary theory. However, even though Darwin did indeed establish the idea of natural selection by 1839, it took him many more years to understand it in detail and to be able to provide a coherent and well-grounded account of how it occurs. After reading Malthus, Darwin wrote in his Notebook D, quoted above, that adapted structures are already formed and are forced into the gaps existing in the economy of nature, or in gaps that occur when other, weaker organisms die out. But, in this view, the adapted structures already exist independent of the local environment. This idea is compatible with the existence of a creator of these forms. Therefore, Darwin's initial view was that natural selection produced perfect adaptation, that is, structures already formed and not ones depending on and relative to the environment.[16]

Even though Darwin's theory was not complete when he read Malthus, and would change in the years to come, its development was contingent *upon* Darwin's reading of Malthus, because the latter's views provided a crucial element for strengthening the analogy between artificial and natural selection. Eventually, natural selection was not the only argument in *On the Origin of Species*, but it was definitely a central one. Furthermore, Darwin's reading of Malthus was contingent *per se*. The works of Malthus were widely read at the time; therefore, reading Malthus in general would not be that unpredictable. But reading Malthus at the particular time that Darwin was reflecting about the importance of artificial selection and thinking about how to establish the analogy between that and natural selection was contingent *per se*, and crucial for what happened next. This is why Darwin's reading of Malthus in 1838 is a **turning point**. It should be noted that reading Malthus at that point was probably not that accidental, as one might infer from Darwin's statement above that he happened to read Malthus for amusement. Rather it seems that Darwin was already becoming interested in statistics and variation, and in his quest for answers he ended up reading Malthus. Through his brother Erasmus he also knew Harriet Martineau, who was a famous proponent of Malthusian views. And the works of Malthus were given particular attention in the writings of Paley, Lyell, and others whom Darwin had already read.[17] But reading Malthus in September 1838 was contingent *per se*, and the further development of Darwin's theory was contingent *upon* it.

The importance of Malthus was also highlighted by Darwin in *On the Origin of Species*:

> In the next chapter the Struggle for Existence amongst all organic beings throughout the world, which inevitably follows from their high geometrical powers of increase, will be treated of. This is the doctrine of Malthus, applied to the whole animal and vegetable kingdoms. As many more individuals of each species are born than can possibly survive; and as, consequently, there is a frequently recurring struggle for existence, it follows that any being, if it vary however slightly in any manner profitable to itself, under the complex and sometimes varying conditions of life, will have a better chance of surviving, and thus be naturally selected. From the strong principle of inheritance, any selected variety will tend to propagate its new and modified form.[18]

So what had Darwin achieved by March 1839? He had come up with the process of natural selection as the main one guiding adaptation. But, as explained, at that point Darwin thought that adaptation was perfect and that therefore there was one best possible form for any given set of conditions. But the theory was not yet complete. In effect, Darwin's theory involved two central ideas, perfectly summarized in his phrase "descent with modification." "Descent" refers to the idea of common descent, that is, that species share common ancestors. "Modification" refers to the idea of transmutation, that is, that species change. By 1839, Darwin was convinced that species change, and he had come up with a process that could account for this change: natural selection. However, certain elements of the theory that we read in *On the Origin of Species* were missing. Darwin was keen to continue his study of nature in order to further develop his theory. And so he did.

Chapter 11

LIKE CONFESSING A MURDER

As described in the previous chapter, Darwin had developed some core concepts of his theory by March 1839. However, he did not immediately proceed to publication and in fact it took him another twenty years before he did so. What was the reason for this? There are two important reasons, which are distinct and interrelated at the same time. The first is that Darwin was aware of the implications of his theory for the received view of the special creation of species. The idea of evolution was of course in the air by that time, but mostly at a speculative level and associated with materialist and other philosophical views. No well-documented theory for the transmutation of species had ever been proposed, which could therefore reject the idea of special creation. As a result, Darwin was seriously concerned about the consequences that his theory would have for him, his family, and his scientific status and career. Therefore, he wanted to establish his theory on the best possible foundation, so that it would become widely accepted. The second, related reason is that Darwin was aware that there were missing elements in his theory; that there were important questions to answer and problems to solve before he would feel confident to proceed with publication, even if there were no reactions to anticipate.

Some core elements of Darwin's theory were put in place in Note-book E,[1] which he started writing in October 1838, as soon as he read Malthus. In the following passage, Darwin mentioned together the analogy between artificial and natural selection, and the role of external conditions for the latter.

It is a beautiful part of my theory, that domesticated races of organics are made by precisely same means as species—but latter far more perfectly & infinitely slower. —No domesticated animal is perfectly adapted to external conditions. —(hence great variation in each birth) from man arbitrarily destroying certain forms & not others. —Term variety may be used to gradation of change which gradation shows it to be the effect of a gradation in difference in external conditions,—as in plant up a mountain—In races the differences depend upon inheritance & in species are only ancient & perfectly adapted races.[2]

Throughout this notebook we can see that Darwin was concerned with the relation between varieties and species, and most importantly with the question of whether new varieties could occur in nature.

Varieties are made in two ways—local varieties, when whole mass of species are subjected to same influence, & this would take place from changing country: but grayhound, race-horse & poulter Pidgeon have not been thus produced, but by training, & crossing & keeping breed pure—& so in plants effectually the offspring are picked & not allowed to cross. —Has nature any process analogous—if so she can produce great ends—But how—even if placed on Isld if &c &c—make the difficulty apparent by cross-questioning—Here give my theory. —excellently true theory.[3]

At the same time, in this very same notebook Darwin mentioned several difficulties and possible objections to his theory. Two important ones were related to how much variation was actually possible to exist in nature, and to the fact that several types seemed to remain constant—which was in turn related to variation, as one would expect types to vary over time.

If It may be said that wild animals will vary according to my Malthusian views, within certain limits, but beyond them not,—argue against this—analogy will certainly allow variation as much as the difference between pi species,—for instance pidgeons—: then comes question of genera.

It certainly appears that swallows have decreased in numbers, what cause??

Seeing the beautiful seed of a Bull Rush I thought, surely no "fortuitous" growth could have produced these innumerable seeds, yet if a seed were produced with infinitesimal advantage it would have better chance of being propagated & so &c.

The greatest difficulty to my theory, is same type of shells in oldest formations: The Cambrian formations do not however, extend round world. Quartz of Falkland. —Old Red Sandstone—Van Diemen's land—Porphyries of Andes.[4]

With all that, Darwin knew that there were questions to answer and problems to solve. In his autobiography, right after describing how Malthus provided him with a key idea to develop his theory, he wrote:

Here, then, I had at last got a theory by which to work; but I was so anxious to avoid prejudice, that I determined not for some time to write even the briefest sketch of it. In June 1842 I first allowed myself the satisfaction of writing a very brief abstract of my theory in pencil in 35 pages; and this was enlarged during the summer of 1844 into one of 230 pages, which I had fairly copied out and still possess.[5]

Why was Darwin anxious to avoid prejudice? Because he was aware of the reactions to previously published theories of evolution. For instance, Lamarck's theory had been severely criticized by Charles Lyell in his *Principles of Geology*. Lyell was a major influence on Darwin, and so he was very concerned about his reaction as well as about that of other eminent men of science. In his book Lyell had written:

We point out to the reader this important chasm in the chain of the evidence, because he might otherwise imagine that we had merely omitted the illustrations for the sake of brevity, but the plain truth is, that there were no examples to be found; and when Lamarck talks "of the efforts of internal sentiment," "the influence of subtle fluids," and the "acts of organization," as causes whereby animals and plants may acquire *new organs*, he gives us names for things, and with a disregard to the strict rules of induction, resorts to fictions, as ideal as the "plastic virtue," and other phantoms of the middle ages.[6]

Darwin strikingly tried to distance himself from Lamarck, in a letter to Joseph Dalton Hooker written in January 11, 1844, in which he also explicitly stated his views about transmutation:

> Besides a general interest about the Southern lands, I have been now ever since my return engaged in a very presumptuous work & which I know no one individual who wd not say a very foolish one. —I was so struck with distribution of Galapagos organisms &c &c & with the character of the American fossil mammifers, &c &c that I determined to collect blindly every sort of fact, which cd bear any way on what are species. —I have read heaps of agricultural & horticultural books, & have never ceased collecting facts—At last gleams of light have come, & I am almost convinced (quite contrary to opinion I started with) that species are not (it is like confessing a murder) immutable. Heaven forfend me from Lamarck nonsense of a "tendency to progression" "adaptations from the slow willing of animals" &c,—but the conclusions I am led to are not widely different from his—though the means of change are wholly so—I think I have found out (here's presumption!) the simple way by which species become exquisitely adapted to various ends.[7]

Species are not immutable, Darwin admitted, and admitting this was like confessing a murder. Hooker replied on January 29, 1844, and only after receiving another letter from Darwin in the meantime, and responded to Darwin's admission by writing: "There may in my opinion have been a series of productions on different spots, & also a gradual change of species. I shall be delighted to hear how you think that this change may have taken place, as no presently conceived opinions satisfy me on the subject."[8]

As already mentioned above, Darwin wrote a thirty-five-page-long sketch of his theory in July 1842, usually referred to as the *1842 Sketch*. This was further expanded during summer 1844 to an essay of 230 pages, usually referred to as the *1844 Essay*. On July 5, 1844, Darwin gave his wife, Emma, this essay along with a letter in which he asked her to publish it in the event of his sudden death. Darwin wrote: "I have just finished my sketch of my species theory. If, as I believe, my theory in time be accepted even by one competent judge, it will be a considerable step in science."[9] Darwin already had an outline at hand, and was concerned, probably because of his occasionally bad health condition, that nobody would find out about his theory if he died prematurely. He therefore wanted to ensure that this would not happen, by asking his wife to have the theory published in that case. But he did not proceed to publication at that point. One reason for this was that Darwin felt that his theory was not yet complete.

However, the event that most likely made Darwin refrain from publication was the publication and the reception of an anonymously published book under the title *Vestiges of the Natural History of Creation*.[10] This book, as readable as a romance, was an ambitious evolutionary epic that combined astronomy, geology, physiology, psychology, anthropology, and theology in a general theory of creation from the formation of the solar system to the emergence of humans. The publication of this book and the public reaction to it seem to have been a turning point itself for the public discussion of evolution. At that time, the publishing and distribution procedure had been significantly transformed and thus books could easily reach an increasing audience. The *Vestiges* attracted public attention, sold fourteen editions and almost forty thousand copies in Britain only, and it was widely discussed in letters, in journals, at parties, and more.[11]

Darwin had access to London research libraries, which were not accessible to most readers at the time, and it was in the library of the British Museum that he read the *Vestiges* in November 1844. He considered that book to be a badly prepared evolutionary manuscript, and he was stunned because, in a sense, the *Vestiges* was a book about the natural origin of species under the effect of material causes and natural laws. Thus, at that point, another author had written publicly about transmutation, attracting attention. Darwin was of course interested in this topic, and at that time was probably intensively thinking about it. In a letter sent to Hooker a few days before he read the *Vestiges*, Darwin wrote:

> . . . but in my most sanguine moments, all I expect, is that I shall be able to show even to sound Naturalists, that there are two sides to the question of the immutability of species;—that facts can be viewed & grouped under the notion of allied species having descended from common stocks. With respect to Books on this subject, I do not know of any systematical ones, except Lamarck's, which is veritable rubbish; but there are plenty, as Lyell, Pritchard &c, on the view of the immutability. Agassiz lately has brought the strongest arguments in favour of immutability. Isidore G. St. Hilaire has written some good Essays, tending towards the muta-bility-side, in the "Suites à Buffon," entitled "Zoolog: Generale." Is it not strange that the author of such a book, as the "Animaux sans Vertebres," shd have written that insects, which never see their eggs, should *will*, (& plants, their seeds) to be of particular forms so as to become attached to particular objects. The other, common (specially Germanic) notion is hardly less absurd, viz that climate, food, &c shd make a Pediculus formed to climb Hair, or woodpecker, to climb trees. —I believe all these absurd views, arise, from no one having, as far as I know, approached the subject on the side of variation under domestication, & having studied all that is known about domestication.[12]

Having briefly reviewed the extant views on the topic, Darwin considered insufficient the pro-transmutation views of Lamarck and others, and believed that his own selection theory could provide the answers. And a few days later, he read the *Vestiges* . . .

A bit later, Darwin received a letter written by Hooker on December 30, 1844, in which Hooker expressed his delight after reading the *Vestiges* because of the multiplicity of facts that it brought together, even though he did not agree with its author's conclusions.[13] On January 7, 1845, Darwin replied to Hooker and expressed his reservations and concerns about the *Vestiges*: "I have, also, read the Vestiges, but have been somewhat less amused at it, than you appear to have been: the writing & arrangement are certainly admirable, but his geology strikes me as bad, & his zoology far worse."[14] Yet, Darwin could not refrain from thinking that if he decided to publish his views, his book would inevitably be compared to the *Vestiges*. He thus followed the public reaction to its publication, imagining the possible reaction to the publication of his own book and evolutionary theory. He realized that he would have to try to distance and distinguish his own theory from the one presented in the *Vestiges*, as well as to prepare—and perhaps address in advance—the possible concerns and critical reactions. A comment he wrote elsewhere is indicative of his own concerns: "The publication of the Vestiges brought out all that cd be said against the theory excellently, if not vehemently." He thus became convinced that the factual grounding of his theory required more work to be done, further, work that he would have to do at the highest possible scientific standards in order to refrain from being called another "Mr. Vestiges."[15]

Probably due to the public reaction to the publication of the *Vestiges*, Darwin refrained from publishing his theory and continued to work on it. He thus managed to add a significant element, that he later considered essential. As he recounted in his autobiography, as soon as he mentioned the writing of the *1842 Sketch* and the *1844 Essay*:

> But at that time I overlooked one problem of great importance; and it is astonishing to me, except on the principle of Columbus and his egg, how I could have overlooked it and its solution. This problem is the tendency in organic beings descended from the same stock to diverge in character as they become modified. That they have diverged greatly is obvious from the manner in which species of all kinds can be classed under genera, genera under families, families under sub-orders, and so forth; and I can remember the very spot in the road, whilst in my carriage, when to my joy the solution occurred to me; and this was long

after I had come to Down. The solution, as I believe, is that the modified offspring of all dominant and increasing forms tend to become adapted to many and highly diversified places in the economy of nature.[16]

Here Darwin referred to the final crucial idea of his theory: the principle of divergence. This is an idea that we find in neither the *Sketch* nor the *Essay*, and it is one that Darwin developed during the 1850s.

In October 1846, Darwin began the study of barnacles, which took him eight years. However, he was often interrupted by illness, which he considered to amount to a total time of two years and which was occasionally so severe that he was unable even to attend his father's funeral. Darwin considered his work on barnacles to be of considerable value in general, and of considerable use in writing about the principles of classification in *On the Origin of Species*.[17] The study of barnacles was especially crucial because thanks to it Darwin managed to answer an important question: whether the variation available in nature was sufficient for natural selection to take place. Variation is a prerequisite for selection; if only few versions of a characteristic are available, it can be the case that none of them will be favored in the environment and thus all members of a population might die out. In contrast, the more versions of a characteristic are available, the more likely it is that one or few of them will be selected, that is, survive and reproduce. Through his study of barnacles, Darwin realized that there was a high degree of variation in all external features. This caused a major shift in his views, and one that is evident between his *1844 Essay* and *On the Origin of Species*. Darwin initially thought that there was little available variation in nature, as perfectly adapted forms could not vary much. Therefore, for new variations to occur, a change in external conditions was required. However, after the barnacles study, he realized that variation was common and natural selection could thus take place anywhere and anytime, leading to transmutation and to the production of better-adapted forms.[18]

At the same time, Darwin became interested in the work of embryologists such as Karl Ernst von Baer, who had suggested that animals developed by progressing from a common pattern to a more specialized one, and Henri Milne-Edwards, who had argued that the diverging paths of development suggested by von Baer corresponded to a natural classification that should have a branching form. Darwin's idea of the principle of divergence was compatible with the views of von Baer and Milne-Edwards, as well as with natural selection. By the time Darwin had completed the barnacles study in 1854, he had conceived of the central idea of the principle of divergence: the ecological division of labor (an idea similar to one proposed by the political economist Adam Smith). This idea could help explain how natural selection could give rise to the various branches of the tree of life. Natural selection could continuously produce better adapted forms by increasing the ecological specialization within groups, the members of which would eventually diverge from the initial form. Thus, the ecological division of labor among animals found in competitive situations would also increase, and natural selection would favor those individuals that were most capable of exploiting new niches. As soon as Darwin came up with the principle of divergence, he also realized that adaptation must be relative and not perfect, as he initially thought: the adaptedness of a species was relative to the adaptedness of other species living in the same area. Therefore, new individuals capable of exploiting new niches were favored by selection, and so there was divergence of form, the direction of which was relative to the environment, as only those who could adapt to it would survive. This shift in Darwin's thinking became evident after he completed the barnacles study, most clearly in 1857.[19]

Let us illustrate the process of divergence, which is essential for understanding how populations and species evolve. Imagine a population of beetles living in a particular environment. Imagine that there exist white, gray, and black beetles, all of which are prey to birds living in the area (see figure 11.1a). Imagine now that there are different-colored areas in that environment (white, gray, and black), in which only particular beetles can conceal themselves. Various subpopulations of beetles, consisting of individuals of all three different colors, live in these differently colored areas.

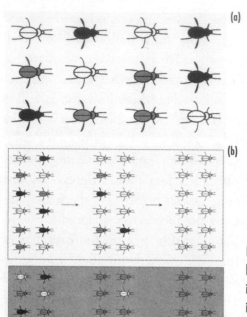

Figure 11.1. If a population of beetles (a) occupies a variable environment, in different parts of which different individuals can survive and reproduce better than others, then it is possible for the various subpopulations of the initial population to gradually diverge through natural selection (b). Eventually, in the long run, these new varieties can differentiate enough to be considered distinct species (c).

Assuming that the white ones are better concealed in the white area, it is possible that birds will spot the gray and black beetles more easily and eat more of those than of the white ones. As a result of this struggle for existence, natural selection can take place and through this the initial subpopulation can evolve to one consisting only of white beetles in the white areas. In the same way, it is possible for the subpopulations living in the gray and the black areas to evolve to ones consisting exclusively of gray and black individuals, respectively. This, again, might happen because the birds living in those areas might spot and eat those beetles that are not well concealed (see figure 11.1b). In this way, the initial population can split into subpopulations with different colors, because the initially similar subpopulations can eventually diverge from one another due to natural selection. In the long run, they might accumulate further changes and evolve to different varieties and perhaps to different species (see figure 11.1c). It should be noted again that the underlying process that can bring about such a change is a natural, unconscious process in which some individuals survive and reproduce more effectively compared to others. As a result, the latter die out without leaving any offspring, whereas the former can pass on their characteristics to the next generation. If this process continues for several generations, one population can evolve to a different one. I must note, though, that the representation in figure 11.1b is oversimplified. The initial population has variation, and so will each of the final ones, independently of whether it will mostly consist of white, gray, or black individuals (in other words, populations cannot consist of individuals of one and the same color). However, in order to avoid confusion, the final populations in figure 11.1c are presented as uniform, even though this is not usually the case because new variation emerges across generations.

The principle of divergence was an important innovation, as Darwin needed to explain how natural selection could give rise to the various branches of the tree of life. In *On the Origin of Species*, he wrote:

Here, then, we see in man's productions the action of what may be called the principle of divergence, causing differences, at first barely appreciable, steadily to increase, and the breeds to diverge in character both from each other and from their common parent. But how, it may be asked, can any analogous principle apply in nature? I believe it can and does apply most efficiently, from the simple circumstance that the more diversified the descendants from any one species become in structure, constitution, and habits, by so much will they be better enabled to seize on many and widely diversified places in the polity of nature, and so be enabled to increase in numbers.[20]

Therefore, it was twenty years after Darwin started his transmutation notebooks that he had a complete theory. In his autobiography he noted that he gained much by his delay in publishing his theory. Even though he first conceived of the theory in 1839 and by 1844 had written an essay on it, he did not publish anything. He hesitated for various reasons, but it seems that the public reaction to the publication of the *Vestiges* was crucial. This is why the publication of *Vestiges* was a **turning point**. The publication of a book on evolution at the time that Darwin hesitated to make his views public was entirely unpredictable. Robert Chambers, who was the author of the *Vestiges*, might have never decided to publish the book, or he could have published it at some other time. In this sense, the publication of the *Vestiges* in November 1844 was contingent *per se*. Most important, the reception of this book and the public reaction toward it enhanced Darwin's hesitance to publish. Being aware of the problems and the flaws in the *Vestiges*, as well as that any book on the topic that he might publish would be compared against that, he decided to work further on his theory and establish it on the best possible foundations. Therefore, the subsequent development of the theory from an essay of 235 pages to a fully developed book-length account was contingent *upon* the public reaction to the publication of the *Vestiges*. Darwin thus continued working on his theory for several years, and around the mid-1850s he had a more complete theory than in 1844. But this is not why he decided to write *On the Origin of Species*. This decision was due to another turning point, as I explain in the next chapter.

Chapter 12

THE MOST STRIKING
COINCIDENCE EVER

I t was only in 1856 that Darwin decided to proceed to the publication of his theory, following the advice of Charles Lyell. In accordance with the standards of the time, he intended to produce a lengthy,
multivolume, detailed, and well-documented work, such as Lyell's *Principles of Geology*. Darwin was now ready to publicly present his theory,
and defend it against criticisms; and he had obtained adequate evidence
to do this in detail in a book that would be called *Natural Selection*.[1] But
his plans changed in June 1858:

> Early in 1856 Lyell advised me to write out my views pretty fully, and I
> began at once to do so on a scale three or four times as extensive as that
> which was afterwards followed in my *Origin of Species*; yet it was only
> an abstract of the materials which I had collected, and I got through
> about half the work on this scale. But my plans were overthrown, for
> early in the summer of 1858 Mr Wallace, who was then in the Malay
> archipelago, sent me an essay *On the Tendency of Varieties to depart
> indefinitely from the Original Type*; and this essay contained exactly the
> same theory as mine.[2]

Alfred Russel Wallace was one of Darwin's numerous correspondents
around the world, who knew that Darwin was interested in the question
of how species originate. In that letter, Wallace had included an outline
of his own answer to this question and asked Darwin to review it and
also send it to Lyell. Whereas Wallace's essay did not employ Darwin's

term "natural selection," it outlined a process of evolutionary divergence of species from preexisting ones due to environmental pressures. Darwin was astonished; at that point he considered Wallace's essay as presenting the same theory with the one on which he had worked for twenty years but had yet to publish. On June 1858, Darwin to wrote to Lyell:

> My dear Lyell
>
> Some year or so ago, you recommended me to read a paper by Wallace in the Annals, which had interested you & as I was writing to him, I knew this would please him much, so I told him. He has to day sent me the enclosed & asked me to forward it to you. It seems to me well worth reading. Your words have come true with a vengeance that I shd. be forestalled. You said this when I explained to you here very briefly my views of "Natural Selection" depending on the Struggle for existence. —I never saw a more striking coincidence. if Wallace had my M.S. sketch written out in 1842 he could not have made a better short abstract! Even his terms now stand as Heads of my Chapters.
>
> Please return me the M.S. which he does not say he wishes me to publish; but I shall of course at once write & offer to send to any Journal. So all my originality, whatever it may amount to, will be smashed. Though my Book, if it will ever have any value, will not be deteriorated; as all the labour consists in the application of the theory.
>
> I hope you will approve of Wallace's sketch, that I may tell him what you say.[3]

Darwin was concerned that all his originality would be smashed. Even though he had spent many years working hard on his theory in order to establish it in the best possible foundations before going public with it, he would eventually be forced to lose or, at best, share the priority with someone else who had come up with the same idea. Darwin's priority was eventually saved, as Lyell and Hooker arranged for a joint presentation at the Linnean Society of both Darwin's and Wallace's papers.

Darwin's paper was based on an abstract of his theory that he had sent to the American botanist Asa Gray as early as September 1857; this letter was also an ideal certification of Darwin's priority in conceiving natural selection. Wallace's paper was essentially based on the essay that he had sent to Darwin. Interestingly, neither of these two were at the presentation at the Linnaean Society! Wallace found out about this event several months later and sent Darwin a letter of approval that arrived early in 1859. Darwin was also unable to participate, because one of his children was seriously ill.

Darwin wrote in his autobiography:

> The circumstances under which I consented at the request of Lyell and Hooker to allow of an extract from my MS., together with a letter to Asa Gray, dated September 5, 1857, to be published at the same time with Wallace's Essay, are given in the *Journal of the Proceedings of the Linnean Society*, 1858, p. 45. I was at first very unwilling to consent, as I thought Mr Wallace might consider my doing so unjustifiable, for I did not then know how generous and noble was his disposition. The extract from my MS. and the letter to Asa Gray had neither been intended for publication, and were badly written. Mr Wallace's essay, on the other hand, was admirably expressed and quite clear. Nevertheless, our joint productions excited very little attention, and the only published notice of them which I can remember was by Professor Haughton of Dublin, whose verdict was that all that was new in them was false, and what was true was old. This shows how necessary it is that any new view should be explained at considerable length in order to arouse public attention.[4]

One might think that Wallace was not treated properly in this case. In the past, it was believed that Darwin had received Wallace's essay before writing to Lyell on June 18, 1858, and that he had intentionally kept it secret for some time in order to decide what to do. The reason that some people thought so was that a letter sent by Wallace to Bates at the same time with the one sent to Darwin arrived in Leicester on June 3,

1858. However, it has been recently concluded that Wallace in fact sent his essay in April 1858, and the postal connections actually indicate that Darwin received the letter at Down House precisely on June 18, 1858.[5] So, Darwin indeed let Lyell immediately know about Wallace's essay. Furthermore, the arrangement made was about a joint presentation, not one of Darwin's paper only while totally ignoring that by Wallace.

After this event, Darwin decided to proceed to publication:

> In September 1858 I set to work by the strong advice of Lyell and Hooker to prepare a volume on the transmutation of species, but was often interrupted by ill-health, and short visits to Dr. Lane's delightful hydropathic establishment at Moor Park. I abstracted the MS. begun on a much larger scale in 1856, and completed the volume on the same reduced scale. It cost me thirteen months and ten days' hard labour. It was published under the title of the *Origin of Species*, in November 1859.[6]

On the Origin of Species was clearly structured and well-developed, providing important evidence for transmutation, but also being incomplete in some aspects. As several critics wrote in reviews of *On the Origin of Species*, Darwin did not manage to provide evidence that transmutation due to natural selection had actually occurred in nature. Natural selection and divergence were Darwin's main principles for how transmutations occurs. As he wrote:

> Owing to this struggle for life, any variation, however slight and from whatever cause proceeding, if it be in any degree profitable to an individual of any species, in its infinitely complex relations to other organic beings and to external nature, will tend to the preservation of that individual, and will generally be inherited by its offspring. The offspring, also, will thus have a better chance of surviving, for, of the many individuals of any species which are periodically born, but a small number can survive. I have called this principle, by which each slight variation, if useful, is preserved, by the term of Natural Selection, in order to

mark its relation to man's power of selection. We have seen that man by selection can certainly produce great results, and can adapt organic beings to his own uses, through the accumulation of slight but useful variations, given to him by the hand of Nature. But Natural Selection, as we shall hereafter see, is a power incessantly ready for action, and is as immeasurably superior to man's feeble efforts, as the works of Nature are to those of Art.[7]

The principle of divergence could explain how natural selection could give rise to the various branches of the tree of life. Darwin wrote:

Here, then, we see in man's productions the action of what may be called the principle of divergence, causing differences, at first barely appreciable, steadily to increase, and the breeds to diverge in character both from each other and from their common parent. But how, it may be asked, can any analogous principle apply in nature? I believe it can and does apply most efficiently, from the simple circumstance that the more diversified the descendants from any one species become in structure, constitution, and habits, by so much will they be better enabled to seize on many and widely diversified places in the polity of nature, and so be enabled to increase in numbers.[8]

These two principles provided the basis for a natural explanation for the origin of species. The main evidence for this, according to Darwin, came from the study of biogeography, that is, from the distribution of species in the various environments:

We are thus brought to the question which has been largely discussed by naturalists, namely, whether species have been created at one or more points of the earth's surface. Undoubtedly there are very many cases of extreme difficulty, in understanding how the same species could possibly have migrated from some one point to the several distant and isolated points, where now found. Nevertheless the simplicity of the view that each species was first produced within a single region captivates

the mind. He who rejects it, rejects the *vera causa* of ordinary generation with subsequent migration, and calls in the agency of a miracle.
... But if the same species can be produced at two separate points, why
do we not find a single mammal common to Europe and Australia or
South America? The conditions of life are nearly the same, so that a
multitude of European animals and plants have become naturalised
in America and Australia; and some of the aboriginal plants are iden-
tically the same at these distant points of the northern and southern
hemispheres? The answer, as I believe, is, that mammals have not been
able to migrate, whereas some plants, from their varied means of dis-
persal, have migrated across the vast and broken interspace. The great
and striking influence which barriers of every kind have had on distribution, is intelligible only on the view that the great majority of
species have been produced on one side alone, and have not been able
to migrate to the other side.[9]

At this point you might wonder how much of a coincidence it was
that Darwin and Wallace came up with the same (actually, *similar*, but
more about this in a while) theory. The answer is that both Darwin and
Wallace were British and therefore familiar with particular characteristics of Victorian era that were crucial for the development of the theory
of natural selection: natural theology was a popular tradition within the
Anglican church that put emphasis on the idea of adaptation; natural
history was very popular in Victorian England, being considered both
as an amusement and as a science, and many people were interested and
actively involved in it; animal and plant breeding was a form of Victorian technology widely practiced at the time, which was also extremely
popular; political economists such as Thomas Malthus and Adam Smith
had developed their theories in the English context, and thus the ideas
of struggle for existence and of division of labor were well known; the
industrialization and imperialism of the British Empire allowed the use
of British ships both to travel around the world in order to do natural
history in remote places and to correspond with other people living there.

Therefore, it is no coincidence at all that Darwin and Wallace came up with a similar theory.[10]

Now, I must note that the two theories are similar but not the same—indeed they differ in some crucial aspects. The first difference is that Wallace did not accept the analogy between artificial and natural selection, which as we saw was central in Darwin's theory. Wallace believed that changes in domesticated organisms were never permanent, as many of these would not support the survival of their bearers in nature. The second difference is the level at which the struggle occurs. Both Darwin and Wallace understood that the struggle for existence occurs among individuals. However, whereas this was fundamental for Darwin, for Wallace it had secondary importance. For Wallace, selection mostly took place at the level of groups, not individuals. A third important difference was that in later years Wallace became a spiritualist and did not accept evolutionary explanations for humans. Finally, Darwin's theoretical and experimental approach supported the development of a more complete and sophisticated theory than that of Wallace. The differences are significant, and most philosophers and historians agree that there are good reasons that the theory is called "Darwinian" and not "Wallacean" (or something like this). Darwin was undoubtedly the first of the two to come up with the idea of natural selection, but this is not the only—perhaps not even the main—reason that the theory is named after him. The main reason is that he came up with a well-established theory.[11]

But Darwin did not, and perhaps could not, realize the differences between Wallace's theory and his own when he received Wallace's letter in June 1858. He thus rushed to publish a book presenting his theory. To achieve this within a reasonable time frame, he decided to drop the big book he had in mind (*Natural Selection*) and instead to write an abstract of his theory (*On the Origin of Species*). This seems to have been a critical decision:

> Another element in the success of the book was its moderate size; and this I owe to the appearance of Mr Wallace's essay; had I published on the scale in which I began to write in 1856, the book would have been

four or five times as large as the *Origin*, and very few would have had the patience to read it.[12]

With these words, Darwin acknowledged the importance of receiving Wallace's essay. This was a **turning point** for the publication of *On the Origin of Species*. First, the reception of Wallace's essay at all, as well as at that particular time during which Darwin was working on his big *Natural Selection* book, was entirely unpredictable. Wallace might have never decided to send the essay to Darwin, or he could have sent it after his book was published. In this sense, the reception of the Wallace essay by Darwin on June 18, 1858, was contingent *per se*. Then, and this is the most important point acknowledged by Darwin himself in the quotation above, the reception of this letter influenced the format in which Darwin's theory was published. As Darwin noted, few people would have the patience to read the long, multivolume treatise he had previously thought was appropriate, whereas a lot more read the shorter (yet still about five-hundred pages long!) *Origin of Species*. Therefore, the subsequent publication and reception of *On the Origin of Species* was contingent *upon* the reception of Wallace's letter.

Wallace wrote that Darwin was, among all men living at the time, best fitted for the great work he had undertaken and accomplished. This is true in a sense because Darwin indeed ended up having a combination of experiences, knowledge, and skills upon which he drew for writing *On the Origin of Species*. But, as we have also seen in the earlier chapters in this part of the book, this was also contingent upon particular critical events that in turn were contingent *per se*. Darwin's theory was published in the form that it was published and with the arguments that it contained neither because Darwin was destined in any way to do so, nor because that was the best possible way to do this, but because of particular turning points that provided him with the experiences, knowledge, and skills that he had and that made him take the particular decisions that he took at those particular times. We cannot really tell what would have happened had Darwin not gone on the *Beagle* voyage, had he died during it, had he not read Malthus in September 1838, had he published his *1844 Essay* despite the likely reactions, or had he never received the Wallace's June 1858 letter and then proceeded with working on *Natural Selection* (see figure 12.1). Educated guesses can of course be made, but we cannot know for sure. Nevertheless, it is conceivable that Darwin could have published his views in some the format of the *Sketch*, the *Essay*, or *Natural Selection*. What is certain is that *On the Origin of Species* was never meant to be written; it was due to particular turning points. The rest is history.

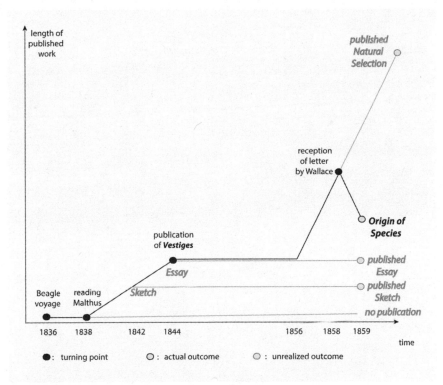

Figure 12.1. The turning points leading to the publication of *On the Origin of Species*, and the related possible but unrealized outcomes (essay and book sizes not in scale).

Part 4

TURNING POINTS IN
HUMAN EVOLUTION

Chapter 13

THOSE TWO FUSED CHROMOSOMES

In chapter 4, I described the views of IDC proponents who consider humans, and the biological world more broadly, as the creation of an intelligent designer. To them, it seems unintelligible that the complexity we see in organisms of all kinds is the outcome of unconscious natural processes. They thus prefer to consider the complex organs and other structures they see as being the outcome of conscious design. Charles Darwin wrote in his autobiography:

> Another source of conviction in the existence of God, connected with the reason and not with the feelings, impresses me as having much more weight. This follows from the extreme difficulty or rather impossibility of conceiving this immense and wonderful universe, including man with his capacity of looking far backwards and far into futurity, as the result of blind chance or necessity. When thus reflecting I feel compelled to look to a First Cause having an intelligent mind in some degree analogous to that of man; and I deserve to be called a Theist. This conclusion was strong in my mind about the time, as far as I can remember, when I wrote the *Origin of Species*; and it is since that time that it has very gradually with many fluctuations become weaker.[1]

Darwin initially found attractive the belief in a first cause, an intelligent mind that was responsible for the immense and wonderful universe. This idea made more sense than accepting that the world was the outcome of chance and necessity. However, he gradually changed his mind. Why?

Because natural evolutionary processes could suffice to explain the observed outcomes. One of these were the similarities between humans and the apes that were so many and so evident. As Darwin wrote in his introduction to *The Descent of Man*:

> The sole object of this work is to consider, firstly, whether man, like every other species, is descended from some pre-existing form; secondly, the manner of his development; and thirdly, the value of the differences between the so-called races of man. As I shall confine myself to these points, it will not be necessary to describe in detail the differences between the several races—an enormous subject which has been fully discussed in many valuable works. The high antiquity of man has recently been demonstrated by the labours of a host of eminent men, beginning with M. Boucher de Perthes; and this is the indispensable basis for understanding his origin. I shall, therefore, take this conclusion for granted, and may refer my readers to the admirable treatises of Sir Charles Lyell, Sir John Lubbock, and others. Nor shall I have occasion to do more than to allude to the amount of difference between man and the anthropomorphous apes; for Prof. Huxley, in the opinion of most competent judges, has conclusively shown that in every single visible character man differs less from the higher apes than these do from the lower members of the same order of Primates.[2]

The skeletal similarities between humans and apes are more than obvious to those who want to see them. DNA analyses have also confirmed our closeness to the apes, with chimpanzees being our closest relatives—in fact, they are genetically closer to us than they are to gorillas (see figure 13.1). There are different estimates about when the evolutionary split between chimpanzees and humans took place, but we can roughly say that we separated from chimpanzees around 5 million years ago, whereas our common ancestors separated from the gorillas around 7 million years ago. In other words, the common ancestor of gorillas and chimpanzees is older than the common ancestor of chimpanzees and humans. This is like saying —very roughly—that chimpanzees and ourselves have the same grandfather and thus are first cousins, whereas gorillas and ourselves have the same great-grandfather and are thus second cousins.[3]

Before we proceed, it is important to make a crucial distinction between two often conflated concepts: ancestry and identity. Gorillas and chimpanzees, our closest relatives, are apes. This entails that we have an ape ancestry, but not that we are apes ourselves (the terms *apes* refers to chimpanzees, gorillas, and orangutans; the term *primates* includes apes, humans, as well as monkeys such as baboons). This distinction is important because stating that we are apes overlooks several of the significant, distinct features of the *Homo* lineage, some of which are discussed in the next chapters. The comparison between humans, chimpanzees, and gorillas at the DNA level cannot tell much about the current differences among us; indeed, scientists were puzzled for a long time how two species that are so similar at the level of DNA, with shared genes accounting for all but a little bit less than 100 percent, are so different in their characteristics. The comparison between humans, chimpanzees, and gorillas at the DNA level can be informative about our ancestry, that is, about where we come from, but not about what we are. Analogously, that we have descended from fish entails that we have a fish ancestry, but not that we are fish. Knowing that we have descended from fish definitely tells us a lot about our ancestry, but this is not the same as our identity. It is more accurate to think that we come from apes, and we are better described as "ex-apes." We are not anything special, but we are unique on this planet. We have an ape ancestry, but we have evolved some significantly different features from our ape ancestors. In this and the following chapters, I explain how we evolved from that point.[4]

The big question then is: what caused the separation of our lineage from that of our close relatives? This is not a question that can be given a straightforward answer. Indeed, several factors may have caused this separation, either at the same time or consecutively. What we can do is search for traces of the effects of such factors in what we see today. This is the

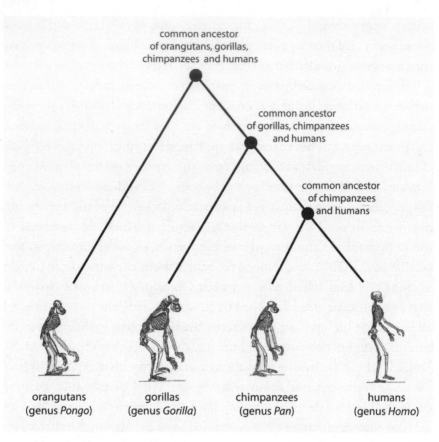

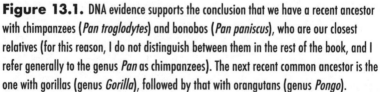

Figure 13.1. DNA evidence supports the conclusion that we have a recent ancestor with chimpanzees (*Pan troglodytes*) and bonobos (*Pan paniscus*), who are our closest relatives (for this reason, I do not distinguish between them in the rest of the book, and I refer generally to the genus *Pan* as chimpanzees). The next recent common ancestor is the one with gorillas (genus *Gorilla*), followed by that with orangutans (genus *Pongo*).

method applied more generally for the study of evolution, which has a distinct historical character. To explain this, let me use a simple analogy. Imagine that you return home and find that a window is broken, and pieces of glass are dispersed on the floor. At that point, you cannot know exactly what happened. Why did the window break? Perhaps you left it open and the wind made it slam closed and break? Perhaps someone from inside or outside the house threw something at it? You do not remember leaving the window open when you left, but it is still a possibility. However, when you look outside, you see some kids who live in the neighborhood playing with a ball. You then think that perhaps it was them who threw a ball at your window and broke it. But what kind of ball could that be? Of what size? Of what material? And with what speed did it hit the window? A baseball thrown at high speed at the window can break it. But a table tennis ball cannot have such an effect. In fact, any ball larger than a baseball could have broken the window. A basketball is heavier than a football, but the latter can be propelled at a higher speed (e.g., kids can kick it harder than they can likely throw it). Eventually, you start looking inside your house for a ball of any of these kinds. If you do not find any ball, you will never know what happened. But if you find one, you can infer that those kids playing outside threw it. And if you find a name on it, or if the parents confirm that it is indeed their child's ball, then you will know how your window broke.

Of course, all of the necessary information might not be available. For instance, if you find a ball but there is no name on it, or if nobody confirms whose it is, you will not know who is responsible for breaking your window. Perhaps a different group of kids were playing there earlier. Perhaps another kid, not the one who owned the ball, threw it at your window. Nevertheless, the important point here is that it is possible to sufficiently explain why your window broke, even if you were not there to observe the event. Finding dispersed pieces of glass and a ball on the floor are traces of the effect that can be sufficient for explaining—if you find them. Note, though, that if you do find this evidence, you do not need to have every single piece of glass nor do you need to know what the weight of the ball is, what exactly it is made of, what its speed was in order to explain why the window was broken. The same principles apply to the study of evolutionary history. We may not have each and every piece of evidence on hand, but with sufficient evidence we can sort out a plausible answer to our questions.[5]

The problem with the study of evolution is that such traces are not always available. Researchers actually use two kinds of traces: fossils and DNA. Fossils are the remains of the hard parts (usually skeletons) of organisms that have been preserved, maintaining the organisms' original shapes. However, many organisms with skeletons have not been fossilized, and many more—such as most invertebrates—cannot be fossilized. But when they are available, fossils are very useful because they provide very important information. One of these is about the era in which the fossilized organisms lived, as it is possible to date the rocks is which the fossils are found. Besides fossils, the other important source of traces is DNA. As explained in chapters 6, 7 and 8, it is possible for mistakes to occur that result in changes, called mutations. These mistakes can take place either during DNA replication and result in changes in the sequence of DNA, or during meiosis or mitosis and result in changes in the structure or number of chromosomes. These mistakes can be lethal for the organism in which they occur; however, some of them may not affect survival and thus may be retained. Thus, the DNA of organisms can have a record of the changes that occurred during evolution, which makes possible to trace the evolution of a certain lineage by studying the DNA differences among species. For instance, the differences between the human and the chimpanzee genomes have been estimated to be approximately 4 percent.[6] Understanding what these differences between the two genomes entail for the characteristics of each species can provide useful insights for understanding human evolution.[7]

One kind of evidence that comparisons between the genetic material of humans and chimpanzees can yield is useful for understanding how these species diverged and for identifying turning points in our evolution. One of these became evident in the early 1980s, when researchers made a comparative study of human, chimpanzee, gorilla, and orangutan chromosomes. Humans have 23 pairs of chromosomes whereas the others have 24. When human, chimpanzee, gorilla, and orangutan chromosomes were colored using a specialized technique that produced bands of different shades on them, and then were compared to one another, a general homology of chromosomal bands in these species was found. This practically means that for each chromosome of each species there was another chromosome in the other species that exhibited the same pattern of bands (imagine comparing barcodes and finding them to be very similar). This not only clearly suggested a common ancestry for these species but also provided clues for the exact relationships among them, indicating that chimpanzees and humans are the closest relatives, and that they share a common ancestor with gorillas. All three in turn shared a common ancestor with orangutans. One of the most striking differences observed was that two individual chromosomes that exist in chimpanzees, gorillas, and orangutans were very similar to the two arms of human chromosome 2. This suggested that a fusion of two chromosomes that existed in the common ancestor resulted in a single, larger chromosome in humans.[8]

To better understand what happened, let us clarify some terms and explain chromosome structure in more detail. In chapter 5 I explained that DNA is packed with proteins, forming a condensed structure called chromatin that during cell division is further condensed to chromosomes. I have also explained that DNA replication and chromatin condensation result in chromosomes consisting of two sister chromatids. Let's break this down further now. Each chromatid has one centromere and two telomeres. The centromere holds the two sister chromatids together and attaches the chromosomes to the appropriate proteins in the cell that make possible the separation of sister chromatids during cell division. The telomeres form special caps at each chromosome end, marking the chromosome's end (see figure 13.2). Overall, the chromatin in centromeres and telomeres is more condensed than in the rest of the chromosome. Because of this and other differences, these parts of chromosomes can be distinguished from each other and from the rest of the chromosome. There are four different kinds of chromosomes, depending on the position of the centromere: metacentric, sub-metacentric, acrocentric, or telocentric. Let's look at the differences among these four types of chromosomes: When the centromere is located around the middle of a chromosome, thus dividing it into two arms that are approximately equal in length, the chromosome is called metacentric. When these arms are not equal, the chromosome is called sub-metacentric (chromosome A in figure 13.2). When the centromere is located near one of the ends of the chromosome, the latter is called acrocentric (chromosome B in figure 13.2). Finally, when the telomere is upon the end of the chromosome, the latter is called telocentric.

What seems to have happened that likely paved the way for our evolutionary split from chimpanzees was that two acrocentric chromosomes in an ancestor of humans fused together, and the centromere of one of them was subsequently inactivated, resulting in a larger sub-metacentric chromosome. As a result of this fusion, sequences that used to be telomeric ones in the ancestral chromosomes, that is, located at their ends, are now located approximately in the middle of human chromosome 2. There actually exist two arrays of degenerate telomere sequences, one opposite to the other, which indicates that chromosome 2 resulted from the telomere-telomere fusion of two smaller chromosomes (see figure 13.3). Therefore, even though humans have 46 chromosomes, whereas chimpanzees, gorillas, and orangutans have 48, the difference is not due to the kind of chromosomes these species have but rather due to how these chromosomes are structured. This means that we have one fewer chromosome just because we have two chromosomes similar to those of our relatives fused together (forming our chromosome 2). Because the fused chromosome is unique to humans, it is assumed that the fusion occurred relatively soon after the human-chimpanzee split, probably around 4 million years ago.[9]

In which cell could this fusion have occurred? It is possible that it occurred in the reproductive cells of one of our ancestors, who like the chimpanzees and the gorillas had 48 chromosomes. The fusion could have happened during the meiotic division leading to the production of an ovum that ended up having 23 chromosomes. When that ovum

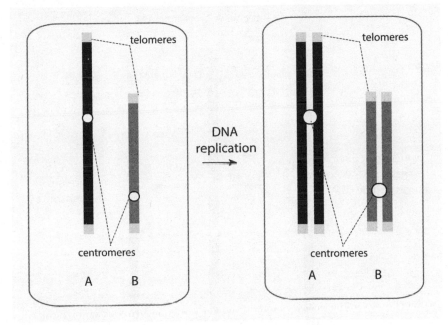

Figure 13.2. The positions of centromeres and telomeres in two sample chromosomes.

of that female individual fused with a spermatozoon with 24 chromosomes, the result was an embryo with 47 chromosomes. Assuming that this new individual was a male, he would have a peculiar chromosome constitution, as the fused chromosomes of his mother would form a chromosome pair with the respective independent chromosomes of his father. However, this person was most likely healthy, as he did not lack any genetic material. All the genetic material included in the reproductive cells of his parents was there; it was just arranged differently (see figure 13.4). If you think that such a phenomenon is highly unlikely, it is not. In fact, it is similar to a "rare" type of chromosomal rearrangement known as Robertsonian translocation (found in one of every one thousand newborns). This translocation results from the fusion of two acrocentric chromosomes; in humans, chromosomes 13, 14, 15, 21, and 22 are acrocentric, and it is possible for

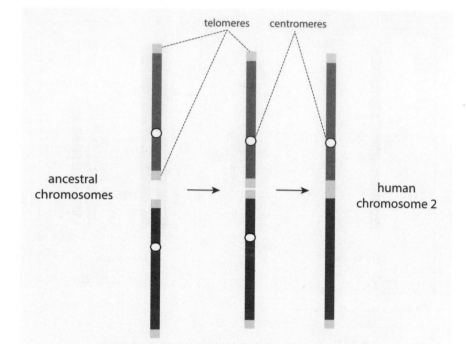

Figure 13.3. The (degenerate) telomeric sequences found in the middle of human chromosome 2 and the correspondence between its arms and certain smaller chromosomes of chimpanzees, gorillas, and orangutans are best explained by the fusion of two ancestral chromosomes that together formed human chromosome 2. This fusion took place between the telomeres of the two chromosomes that are now found in the middle of human chromosome 2.

any of them to be involved in a Robertsonian translocation. When a Robertsonian translocation happens, the result is a metacentric chromosome consisting of the long arms of the two chromosomes that fused. Theoretically, all possible combinations of chromosomes 13, 14, 15, 21, and 22 are possible. However, in most cases it is two different chromosomes that fuse, with the most common being that between chromosomes 13 and 14. This can happen during the first meiotic division, which, as I have already mentioned, can last for several years (about ten to fifty years) during a woman's life. This provides adequate time for such mistakes to occur.[10]

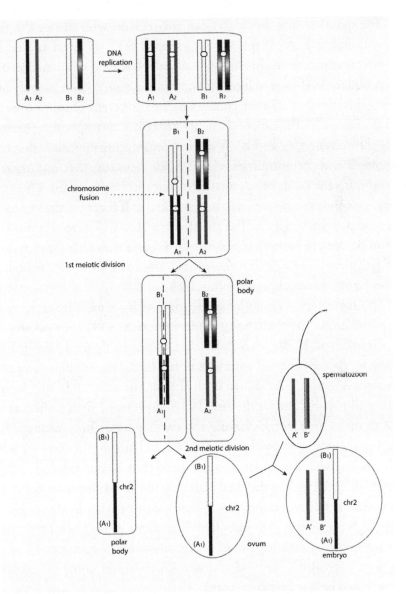

Figure 13.4. A fusion of two acrocentric chromosomes can produce a larger submetacentric chromosome. As a result, the new individual will have one fewer chromosome than his parents each have. Yet he will also likely have a complete set of genetic material.

The question now becomes: can this person who carries the fused chromosome 2 have offspring? To explore this, we need to see how meiosis can proceed in this individual and what kind of spermatozoa he can produce. As shown in figure 13.4, this individual has a long chromosome (chromosome 2) that consists of two different chromosomes, A_1 and B_1. For each of these constituents, there is a corresponding homologous chromosome, A' and B'. What happens during meiosis is that chromosome 2 and chromosomes A' and B' all meet together, and there are several different kinds of spermatozoa that can be produced. One possibility is for the chromosomes to be separated as if there were no fusion. In this case, approximately half of the spermatozoa will have chromosomes A' and B', derived from that person's father, and the other (approximate) half will have the large chromosome 2 resulting from the fusion and derived from the mother (see figure 13.5).

This male person could have offspring with a female from the rest of the population. These offspring could have either 48 chromosomes like their mother or 47 like their father, as shown in figure 13.6a (for simplicity, let's accept that all members of the initial population carry chromosomes A' and B'). As shown in the figure, the offspring could be both males and females having the fused chromosome of their father, as well as both males and females having the two separate chromosomes. Now imagine that these individuals lived in a polygamous society in which males mated with several females. It would thus be possible to have even more male and female individuals carrying the fused chromosome 2. And when these would mate, they could have offspring with 48 chromosomes, offspring with 47 chromosomes like themselves, as well as offspring with two chromosomes 2 and overall 46 chromosomes (see figure 13.6b). In this way, individuals with 46 chromosomes would come to existence. These would be our first ancestors.[11]

Beyond this point a lot could have happened. It seems that the separation between humans and chimpanzees was not abrupt at all, and that before the two species were permanently separated there was extensive

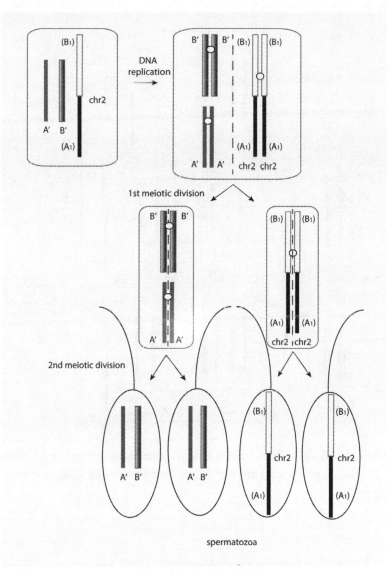

Figure 13.5. The male person who carries the chromosome resulting from the fusion can be genetically normal, as he does not lack any genetic material, and he can produce spermatozoa that are chromosomally normal at the quantitative level (but not qualitatively, as A_1 and B_1 are fused, forming chromosome 2).

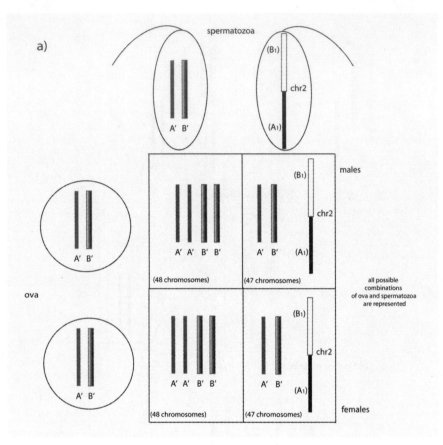

Figure 13.6. (a) The male person having the chromosome resulting from the fusion could have offspring who also have it (males and females are depicted separately in this figure).

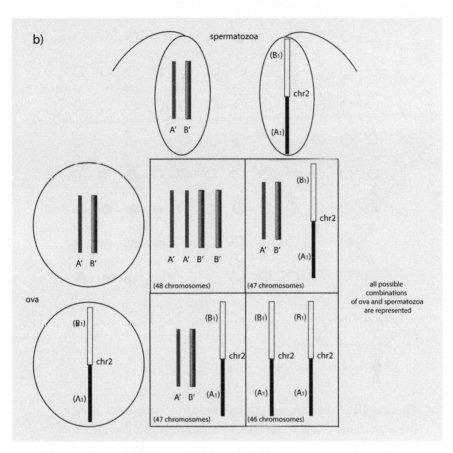

Figure 13.6. (b) When two individuals carrying chromosome 2 mated, it would be possible to have a new type of individual with two chromosomes 2 and overall 46 chromosomes (the genotypes of offspring that could be either male or female are presented in this figure).

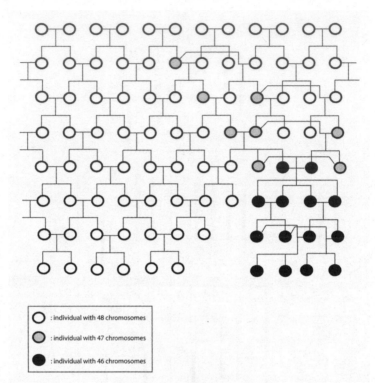

○ : individual with 48 chromosomes

◐ : individual with 47 chromosomes

● : individual with 46 chromosomes

Figure 13.7. The emergence of an individual with 47 chromosomes through a chromosome fusion could be the starting point for the emergence of more like him, as well as of individuals with 46 chromosomes through sexual reproduction. Further genetic changes, which might render the offspring of individuals with different chromosome numbers infertile, might make individuals to be able to have offspring only with others having the same chromosome number. As a result, the initial mixed population could gradually be separated into two subpopulations, one consisting of individuals with 48 chromosomes and another consisting of individuals with 46 chromosomes. These two subpopulations would be reproductively isolated, that is, individuals from one subpopulation could not produce fertile offspring with individuals from the other subpopulation. Further changes in physiology, behavior, and genetics might lead to the emergence of two new species, such as ourselves and the chimpanzees. It should be noted that for the sake of simplicity, the pace of the respective evolutionary events has been exaggerated in this figure.

interbreeding during the early period of the divergence.[12] This means that once the individuals with 46 chromosomes emerged, they continued to interbreed and have offspring with the individuals having 47 or 48 chromosomes. But it is conceivable that at a certain stage further molecular and/or chromosomal changes could have created a reproductive barrier that precluded further interbreeding, as the offspring of these individuals could be infertile. This could happen, for instance, if chromosome 2 and chromosomes A' and B', which in figure 13.5 are presented as being homologous and "meeting" each other during meiosis, were differentiated to the extent that they would no longer "meet." If this happened, meiosis would not take place properly; therefore, there would be no reproductive cells and the offspring would thus be infertile. In this way, individuals with 47 chromosomes might gradually disappear, and there would be only individuals with either 48 or 46 chromosomes. Figure 13.7 provides an oversimplified illustration of this process.

What I have presented so far is a plausible scenario for how chimpanzees and humans diverged from our common ancestor. All of the available evidence is best explained by this scenario, even though there still exist details that we do not understand. I must note at this point that scientists are never 100 percent certain about all of their conclusions. Their usual approach to data is to infer the best explanation from them, one that most clearly explains the available data and provides the most complete understanding.[13] Anti-evolutionists often claim that humans cannot have emerged by chance and random processes. But this is not what evolutionary biologists claim; evolutionary biologists suggest that random changes, such as the chromosome fusion, provide material for selection to take place in nature. As I have shown in this chapter, the available evidence includes traces of differences at the chromosomal level between humans and chimpanzees, gorillas, and orangutans: two chromosomes of these species correspond to a single human chromosome that has the features that one would expect after a chromosome fusion. This can be explained as the result of chromosomal rearrangements that are not the product of some creator's imagination but that are similar to ones observed today, such as the Robertsonian translocation. Evolutionary biologists provide a plausible and sufficient natural explanation for what we observe.

Therefore, the chromosome fusion that likely occurred in our ancestors about 4 million years ago was a **turning point** in human evolution. This fusion was an event contingent *per se*, as chromosomal rearrangements like this one are totally unpredictable. Whether or not the fusion shown in figure 13.4 will occur between particular chromosomes cannot be predicted in advance. In many cases, nothing like this will happen. However, even if the fusion that took place between those two chromosomes in our ancestor was not very likely, it could have happened like the Robertsonian translocations observed today, with a frequency of 1/1,000. Once it occurred, this change could have paved the way for the divergence of the initial population into two subpopulations and eventually to two species with different chromosome numbers, as shown in figure 13.7. In this sense, subsequent events were contingent *upon* the fusion of the two chromosomes.

If such a reproductive isolation of our ancestors from their close relatives took place, then we should expect to see particular characteristics restricted to our own species, and not shared by our own close relatives (such as the chimpanzees). The reason for this is that once our human ancestors became reproductively isolated, the genetic predisposition for particular characteristics could be transmitted to the offspring of our ancestors but not to those of the ancestors of the chimpanzees. The reason for this is that there was no interbreeding and no co-contribution of genetic material to produce offspring, or that (if there were offspring) the offspring themselves were not fertile and could not reproduce further. This is why the characteristics that I am presenting in the next chapters are exclusive characteristics of present-day humans.

Chapter 14

STANDING UP, WALKING UPRIGHT

One of our unique features is our ability to stand upright and walk on our legs without any need to use our arms. Even though several apes can stand upon their legs too, none of them can walk in the way we do; we are the only biped, that is, the only primate that uses only its legs for walking. Charles Darwin wrote in *The Descent of Man*:

> Man alone has become a biped; and we can, I think, partly see how he has come to assume his erect attitude, which forms one of the most conspicuous differences between him and his nearest allies. Man could not have attained his present dominant position in the world without the use of his hands which are so admirably adapted to act in obedience to his will. As Sir C. Bell insists "the hand supplies all instruments, and by its correspondence with the intellect gives him universal dominion." But the hands and arms could hardly have become perfect enough to have manufactured weapons, or to have hurled stones and spears with a true aim, as long as they were habitually used for locomotion and for supporting the whole weight of the body, or as long as they were especially well adapted, as previously remarked, for climbing trees. Such rough treatment would also have blunted the sense of touch, on which their delicate use largely depends. From these causes alone it would have been an advantage to man to have become a biped.[1]

Darwin noted that bipedalism likely provided an evolutionary advantage to our ancestors, who could henceforth use the upper part of their body, and particular their forelimbs, for other activities than simply supporting the weight of their body and moving around. Before we consider what these advantages could be, let's see in a bit more detail what it means to be a bipedal organism.

Our "cousins," the chimpanzees and the gorillas, walk and run in a way that looks peculiar to us, called knuckle-walking. In doing this, they use their forelimbs to support their weight, holding their fingers in a way that allows them to press on the ground through the knuckles. In contrast, we are able to stand, walk, and even run in an upright position. Our bipedalism seems to be the outcome of particular skeletal changes that occurred in our lineage and that are not found in our closest relatives. One such difference from chimpanzees is the shape of the bones that form the upper part of our pelvis. Whereas in apes these bones are tall and face backward, in our case they are short and face sideways. This is very important because it allows the muscles on the sides of our hips to stabilize our upper body over each leg when, in the process of walking, only one of our legs is on the ground. Another such difference is our spine, which has a S-like shape. Whereas the spines of apes have one curve that makes their trunks lean forward, our spines have two curves which together allow for a vertical configuration of the upper body. A third change is the arch of the foot, which is created by the shape of the bones of the foot and of the respective ligaments and muscles, and which enable us to push the body upward and forward while walking. A fourth related striking difference is the shape of the human foot itself. Among all of the primates, we are the only ones that do not have an opposable big toe, instead we have one that is in line with the other toes. As all toes are relatively short and straight, the human foot is a propulsive platform that supports bipedal walking as well as standing up and supporting the weight of the body. There are several other differences between our skeleton and that of apes, which scientists consider essential for bipedalism. As a result of such differences, we are the only exclusively bipedal organisms among more than two hundred extant primate species.[2]

The study of the available fossils of ancestral hominin[3] species points to the conclusion that the early hominins were occasional bipeds; this means that they walked on two legs occasionally. In contrast, the various individuals of the genus *Australopithecus* could be considered habitual bipeds; this means that they had the habit of walking on two legs, but they also engaged in some kind of arboreal locomotion. However, even within this genus there was diversity, as *Australopithecus afarensis* (the species of the famous Lucy fossil) exhibits a mix of features related to arboreal life and bipedalism, which is nevertheless more human-like than that of *Australopithecus africanus*. Finally, the characteristics of the leg and foot of *Australopithecus sediba* are considered to indicate a form of bipedalism distinct from that of both early *Homo* and other species of *Australopithecus*. In contrast, it seems that only the species of the genus *Homo* are obligate bipeds, that is, after the first few years of life, bipedal locomotion is the only one used. The available fossil evidence overall supports the conclusion that true obligate bipedal locomotion emerged sometime between about 2.5 and 1.8 million years ago, a period also associated with the emergence of the genus *Homo*. By 1.8 million years ago, at least *Homo ergaster* was a fully obligate biped with features of a modern human body. Other species, such as *Homo habilis*, may have had a kind of locomotion behavior transitional between that of *Australopithecus* and of *Homo ergaster*. Finally, the species emerging after *Homo ergaster*, such as *Homo erectus*, *Homo neanderthalensis*, and ourselves—that is, *Homo sapiens*—seem to have been fully obligate bipeds. It should be noted, though, that the recent findings of *Homo floresiensis*, exhibiting a different type of bipedalism than modern humans, and *Homo naledi*, exhibiting features related to arboreal life, further complicate this picture. This is why you should keep in mind that the details are complex, the evidence is incomplete, and the conclusions are far from definitive. However, overall, one can roughly perceive a gradual evolution from occasional to obligate bipedalism, which was nevertheless not linear but quite complex and diverse.[4]

Now the two important questions are: why did bipedalism evolve around 2 million years ago, and what impact did it have on hominin evolution thereafter? Let us consider the first question. There are several different explanations about why bipedalism could have evolved, and in particular what kind of evolutionary advantage it could have conferred. In this chapter I will briefly consider a few of these explanations. But before doing so, it is important to consider that although it is easy to come up with several hypotheses, or even plausible explanations, these have to be supported by the available evidence. Unfortunately, we can study fossils and DNA, but even when these pieces of evidence are complete, they do not give us much information about behaviors and life habits of earlier species. Therefore, there is a lot that is left to inference. In this case, the best we can do is look at the currently available evidence. As our common ancestor was most likely chimpanzee-like rather than human-like, we can compare ourselves to our closest relatives and find the differences between their and our structure and behavior. On the basis of that, any advantages that we have as bipedal animals over the chimpanzees could have been the factor that made the difference so that bipedalism became prevalent through selection in our human ancestors.

The first explanation was suggested by Darwin in the quotation above. As he noted, becoming bipedal sets free the whole upper part of the body and especially the arms. Thus, it would be possible to make weapons and tools that eventually helped our ancestors dominate the world. The idea here is that the more bipedal an individual is, the less this individual needs to use the arms for locomotion. Therefore, the arms are free for other uses, and they could thus have evolved to their present condition that makes us capable to construct tools and weapons and use them for a purpose. However, a big problem with this idea is that whereas bipedalism, at least habitual bipedalism, is evident in *Australopithecus* around 4 million years ago, there is evidence for the use of sharp stones for cutting only around 3 million years ago and for stone tools that were intentionally given their shape only around 1.7 million years ago. Furthermore, the structure of the hand of *Australopithecus* does not indicate that they were any better than chimpanzees in making tools, as their thumb was smaller and weaker than ours.[5] Therefore, it is difficult to maintain that biped-alism evolved because it provided the advantage of toolmaking.

Another explanation is based on sexual selection, which Darwin also advanced in *The Descent of Man*. Broadly put, sexual selection has to do with the selection of characteristics that confer a reproductive advan-tage. These characteristics are selected (naturally) because their bearers seem more attractive than others to the members of the opposite sex, and so they are more likely to mate with them and thus have offspring. This means that these characteristics may not be advantageous for survival, which is a prerequisite for reproduction; rather, these characteristics are advantageous for reproduction itself, with the eventual outcome that—whatever these characteristics are—are passed on to the next generation.[6] An explanation of this kind has been that bipedalism exposes one's sexual organs to the eyes of others, especially when that person has neither fur nor clothes. Women do not exhibit any overt physical signs that they are close to ovulation and thus capable of having offspring; this can be unknown even to themselves, in contrast to the females of other primate species who exhibit physical signs that they are receptive and likely to have offspring. Therefore, it could have been the case that a bipedal male's visible genitalia could have attracted female attention in our ancestors (this could also have been the case for the breasts and the genitalia of the females, which, by being visible, could have similarly attracted the atten-tion of males). If it were the case, as it is for most women today, that it can be unknown when a female is ovulating and can thus get pregnant and have offspring, visual cues such as the genitalia of an individual of the opposite sex could have gradually become more and more impor-tant for engaging in sexual intercourse. Not knowing precisely when they could get pregnant, females could have chosen among those males that reminded them that sex is a prerequisite for having offspring. In this sense, the more bipedal a male was, the more visible his genitalia were, and the more likely it was for females to be attracted to and mate with him. Thus, bipedalism could have become more and more prevalent in the populations of our ancestors.[7] This suggestion is quite speculative but nevertheless points to actual differences between humans and apes.

Another suggestion is that being bipedal provides a thermoregulatory advantage to primates living in an African savannah. The obvious advantage in that case would be that a bipedal organism is less exposed to the solar rays that would be vertical in direction in that area. Bipedal primates have only the top of their heads exposed, whereas quadrupeds have not only their heads but also a large part of their back part of the body exposed as they tilt forward. Therefore, bipedalism allows for a reduced absorption of heat from the sun. At the same time, most of the body would be away from the hot ground, thus increasing exposure to cooling breezes. This and the low relative humidity above any vegetation on the ground would also increase the rate at which sweat can be evaporated from the skin.[8] One issue with this model is that it is based on the assumption of a standing and not a walking primate. Another is that it assumes that early hominins were as hairless as we are today. Insofar as early hominins were hair-covered, they would have serious thermoregulatory problems walking around in hot, sunny environments, and bipedalism could not have made any difference in thermoregulation. This entails that bipedalism could not have evolved because of a thermoregulatory advantage in hair-covered ancestors; but if it had evolved because of another reason, it would cause no problem. However, when our ancestors ended up having the hair and the sweating ability that we humans currently have, it would have been possible for them to have good thermoregulation even in a hot, sunny environment, and even when they were walking, which is an activity that is energy-demanding. This means that bipedalism could have evolved because of a thermoregulatory advantage only if hair loss had evolved before that.[9]

A different explanation for the origin of bipedalism has been that it facilitates locomotion on flexible branches. As already described above, several of our ancestors exhibit a mixture of features, some of which are related to bipedalism and others to arboreal life. Therefore, bipedalism may have been initially selected because it provided an advantage for life in trees, and it was only later that it facilitated life on the ground. This could have been similar to the hand-assisted bipedalism of orangutans that allows them to move on flexible branches that are otherwise too small to access. To explore this, researchers conducted a year-long field study of orangutans, observing 2,811 instances of their ways of moving. The analysis of these observations showed that bipedalism was strongly associated with locomotion on multiple branches of a small diameter, whereas quadrupedalism was associated more with locomotion on single, large branches. In most observations of bipedalism, orangutans extended their arms for stabilization. Locomotion on flexible branches is safer if it is supported from both above and below, and at the same time bipedalism allows a hand to reach for food; therefore, the researchers concluded that bipedal locomotion can confer selective advantages to arboreal apes, such as facilitating stabilization and foraging in parts of trees where branches are more flexible. In this sense, it may have provided an advantage to our arboreal ape-like ancestors, and it was in this way that our bipedalism may have initially evolved. In this sense, humans retained some skeletal features already present in the common ancestor with orangutans, whereas chimpanzees and gorillas diversified toward a different way of locomotion.[10]

Similar observations were made by other researchers who documented 179 instances of bipedal posture in several chimpanzees over a two-year study. In this case, bipedalism was observed only in the trees, and it was a means of standing rather than a means of locomotion, mostly for adult chimpanzees. It was observed within a complex series of behaviors, the vast majority of which occurred when chimpanzees were trying to get fruit from trees. Interestingly, bipedalism was mostly observed on large branches, and actually its frequency and duration were higher, the larger were the branches. Based on these observations, the researchers suggested that the origin of bipedalism in hominids could be related to advantages in foraging on fruit trees. But it should be noted that in this case, it is postural bipedalism and not locomotor bipedalism that it was observed. This means that standing up on two feet might have provided an evolutionary advantage in obtaining food, and therefore features related to upright posture could have been favored by natural selection. According to these researchers, then, standing up therefore preceded and may have been independent from walking upright.[11]

Observations like these seem to support the conclusion that the ability to stand, and eventually walk, upright may have provided an advantage in foraging. The more bipedal an individual was, the more efficient it could be in foraging for and obtaining food, and therefore in surviving and reproducing. But why was this so important? Even if they did not stand upright, couldn't our ancestors still forage in trees? There is evidence that between 10 and 5 million years ago Earth's climate became cooler. As a result of this change, which took place over a long time and with fluctuations, the rainforests of Africa eventually shrunk and the woodlands expanded. Therefore, for those of our ancestors living at the edges of the rainforests, fruit could have been less and less available. This could have forced them to either eat other types of food, such as stems and leaves that are of lower quality than fruit, or travel far to look for food of good quality. It is under such conditions that bipedalism could have provided an advantage in traveling far and carrying food back to the places where they lived. A bipedal individual can look far away and search for food, and also has its arms free to obtain this food and carry it back to the place where it lives. Chimpanzees and gorillas can carry food with one arm, but they still need the other one for knuckle-walking. In contrast, a bipedal individual can use both arms to carry more food than chimps can, and so be more efficient in foraging and feeding.[12]

A related explanation is based on the differences in the consumption of energy between bipedal and quadrupedal walking. The idea behind this is the longstanding hypothesis that bipedalism requires less energy than the knuckle-walking of apes and our ape-like ancestors. As a result, bipedal individuals required less energy for foraging and they therefore came to have an evolutionary advantage over other apes, perhaps because the energy saved could be used for other functions or because they could walk for a longer time compared to others. To test this hypothesis, researchers compared the energetics and biomechanics of locomotion in four humans and five adult chimpanzees, in an attempt to see how anatomy, walking, and energy cost are related. Subjects walked on treadmills while wearing masks that collected expired air, which allowed researchers to measure oxygen consumption. The researchers found that chimpanzee quadrupedal and bipedal walking is approximately four times costlier than human walking. In addition, the variation in cost can be explained by biomechanical differences in anatomy and walking. For instance, it was found that the shorter legs of chimpanzees make them generate greater ground forces at a faster rate than humans, thereby increasing the energy cost required for walking. At the same time, whereas the long arms of chimpanzees facilitate knuckle-walking at a similar speed as humans' upright walking, the walking costs are higher in chimpanzees than in humans. Therefore, one can reasonably infer that if early hominins shared a similar anatomical structure and locomotion behavior with chimpanzees, which is indeed the case, the emergence of the less-costly bipedalism could have been an advantage.[13]

A different explanation for the evolution of bipedalism is that the ability to stand up has allowed humans the use of forearms in fight, providing an advantage to those individuals who did that, which in turn favored the evolution of bipedalism. Standing up on the hind legs and fighting is often observed in various mammals including domesticated cats, lions, tigers, foxes, wolves, dogs, bears, rodents, and primates. This may happen because mammals are able to strike their opponent with greater force from a bipedal posture than from a quadrupedal posture. In addition, it could be the case that mammals could exercise greater force when striking their opponent downward with their forelegs from a bipedal posture than up from a quadrupedal posture. A study aimed at measuring the force and energy produced when humans struck from "quadrupedal" and bipedal postures, as well as when they struck downward and upward from these positions. It was found that for all different types of strike (downward, upward, side, and forward), there was a greater performance from a bipedal than from a quadrupedal posture. In particular, when striking downward and upward from a bipedal posture, the energy released was 44 percent and 47 percent more than the respective quadrupedal posture. Similarly, for side and forward strikes the force exercised from a bipedal posture was 30 percent and 43 percent greater than from the respective quadrupedal postures. It was noted that when participants struck vertically from a bipedal posture, they used approximately the full range of motion of their arm, something that was not possible from a quadrupedal posture. This could explain the differences in the energy released in the respective strikes. At the same time, the greater performance in side and forward directed strikes from bipedal posture could be explained partially with the transfer of energy from the legs and trunk. Such results can form the basis for the conclusion that bipedal posture provides a performance advantage for striking with the forelimbs. The researchers' conclusion in this case was that as the mating processes of great apes involve intense male-male competition and the use of force, fighting from a bipedal posture could have been favored by sexual selection and thus contributed to the evolution of (habitual, in this case) bipedalism in our ancestors.[14] In other words, the more "bipedal" a male was, the more efficient it could be in fighting other males and eventually the more likely it would be to get rid of them and mate with the females.

Where does all of this bring us? We are looking here for the reasons that bipedalism evolved in the human lineage only, and eventually we assess its significance for human evolution. Was the emergence of bipedalism of critical importance in human evolution, and, if so, why? One might go on by conceiving explanations that are more or less supported by the available evidence. As Henry Gee put it in his book *The Accidental Species*, "armchair theorists" of bipedalism can select any piece of evidence or information they want and develop an explanation about why bipedalism evolved. However, "there can be no simple relationship between a proposed cause and a proposed effect. The consequence of one change has an impact on many other traits or adaptations, until the whole body is affected. . . . Bipedality means more than just standing on two legs. It requires the wholesale modification of the body, not all of it very effective."[15] This is a very important point, because it is easy to look at single changes in the foot or the knee or the pelvis, or wherever, and try to use them in order to provide an explanation for how bipedalism evolved. But the problem is that all of these changes support bipedalism, and we cannot know all of the details about which one came first, at what time, and which others followed. We should not forget that populations that evolve due to natural selection evolve due to the differential survival and reproduction of individuals as wholes. So when we look at how hominins became occasional bipeds, habitual bipeds and eventually obligate bipeds, we also need to look at when and how all of the relevant anatomical changes took place.

It should be noted once again that there is no need, and indeed no evidence, for a linear transition. As orangutans and chimpanzees do today, our ancestors could have used a variety of locomotion behaviors and postures, depending on the conditions under which they lived. However, if some individuals happened to have variations in their skeleton (in posture and/or locomotion) that favored bipedalism, these individuals could have been more efficient than others in collecting fruit, and eventually surviving and reproducing. In this way, the respective features could have become more prevalent in subsequent generations. An important point to note is that selection does not take place from a fixed condition where variants A and B are present toward either A or B. Rather, selection takes place when a lot of variation is available. For instance, we think that evolution could have proceeded from a hypothetical ape-like ancestor with four or three lumbar vertebrae (the cylindrical bones of which the lower part of the spine consists) to an early hominin with five lumbar vertebrae. However, chimpanzees with five lumbar vertebrae also exist, even though they are rare.[16] Research in developmental genetics and evolutionary developmental biology supports the conclusion that small molecular changes can produce extensive changes in the body structure. This means that whenever we think about the evolution of bipedalism discussed in this chapter, which is accompanied by various changes in the body structure, we should keep in mind that these changes do not need to have accumulated in a linear manner. Rather, several combinations of these features may have existed, and the evolution of our current form may have been complex and messy.

Finally, I must point out that bipedalism is not exclusively advantageous. In contrast, it could have posed several challenges to our ancestors. For example, bipedal animals cannot run at the same speed as quadruped animals. They are also less stable when running and changing direction than are quadruped animals—just watch a leopard run while chasing and how it changes direction without reducing speed, and wonder if you can do the same. In addition, bipedalism probably makes climbing trees a lot more difficult. Watch a chimpanzee do this, and you will realize that even the most competent and athletic human cannot get even close in climbing so easily and so quickly. Pregnant women also have to carry a lot of weight on two legs, with the eventual consequence of suffering from back pain. To this challenge we could add occasional problems in our ankles, knees, and backs. Therefore, we should not think of bipedalism as an exclusively advantageous posture that appeared and evolved through natural selection. Rather, we should think of it as the outcome of a complex set of anatomical changes that brought about advantages and disadvantages. Eventually, our own presence in this world somehow confirms that the advantages outweighed the disadvantages. But the evolutionary process must have been complex, if not messy.[17]

So how did bipedalism evolve? It seems that there are two hypotheses. The first is that humans had already adopted the upright posture while still living in trees and retained it when they moved to the ground, eventually becoming bipedal. The common ancestor of humans, chimpanzees, and gorillas in this case could have been similar to orangutans, which exhibit what can be called forearm-assisted bipedalism and which could be considered as a precursor to terrestrial bipedalism. If this is the case, then knuckle-walking must have evolved independently in gorillas and chimpanzees, whereas terrestrial bipedalism evolved in humans. Under this hypothesis, in other words, three evolutionary changes are needed (see figure 14.1a). The other hypothesis is that the common ancestor of gorillas, chimpanzees, and humans was a knuckle-walker. In this case, this particular characteristic was retained in the lineages of gorillas and chimpanzees, and humans evolved terrestrial bipedalism (see figure 14.1b). This hypothesis is simpler, because it only assumes two changes instead of three, and indeed the anatomical evidence better supports it. I must note though that all these changes (forearm-assisted bipedalism → knuckle-walking; forearm-assisted bipedalism → bipedalism; knuckle-walking → bipedalism) are complex changes that require several anatomical transformations.[18]

It will nevertheless be very difficult to figure out what exactly turned the evolution of hominins towards bipedalism. Isn't this a problem? Well, there is another way to look at the issue. Rather than looking into potential advantages of bipedal posture and walking, and trying to figure out which one of them was THE key one, we should keep in mind that once bipedalism was adopted, all the potential advantages mentioned above were there.[19] This means that no matter the reason why a more upright posture was initially favored, there were also other potential advantages that might have further favored this change. Therefore, it could have been a minor change that provided an advantage and made more of the "a-bit-more-upright" individuals survive and reproduce better than others, for example, because they were better at foraging in trees. These "a-bit-more-upright" individuals could have then been able to also walk on the ground and explore areas farther away, and find more food. Among them, those who were "even-more-upright" could eventually have survived and reproduced better than others and eventually spent more time on the ground. In this way, they could have exploited further advantages and eventually diverge even more from their arboreal relatives. In short, no matter which advantage favored some initial form of bipedalism, there were several other advantages that could further favor evolution in the direction that leads to us.

Figure 14.1. Two hypotheses for the evolution of bipedalism: (a) The common ancestor had an upright posture and some form of bipedalism, like orangutans. From such forms, evolution drove humans, on one hand, and chimpanzees and gorillas, on the other hand, to different directions; (b) The common ancestor was similar to chimpanzees and gorillas, and it was orangutans and humans who independently diverged toward different forms of bipedalism. The latter hypothesis requires fewer changes and is better supported by the anatomical data.

Bipedalism was therefore the first major transformation toward modern humans. This change was not inevitable but contingent *per se*. The various anatomical changes that are characteristic of bipedal organisms, in whatever order they may have occurred, were contingent *per se*, as it would be impossible to have been predicted in advance. For example, whether or not an organism will have a pelvis with larger or smaller bones facing one or another direction is a change that one cannot predict in advance. Such minor changes can form the basis for further changes that are also unpredictable, for example, smaller upper pelvic bones that allow the muscles on the sides of our hips to stabilize our upper body over each leg when only one leg is on the ground while walking. Such changes are contingent *per se* but also contingent *upon* the changes that have occurred earlier. This in turn can favor other changes that will further favor the evolution of bipedalism. Each of these minor anatomical changes can be considered as a **turning point**, as it is contingent *per se* and subsequent changes will be contingent *upon* it (some will be possible, others will not). Bipedalism itself is also a major turning point. It is certainly contingent *per se*, as one could not have predicted in advance that it would emerge, and that the respective anatomical changes would have taken place. At the same time, it seems that the subsequent evolution of our species was contingent *upon* it. Bipedalism allowed terrestrial life, the use of hands for foraging and later for toolmaking, and other habits that paved the way for the evolution of *Homo sapiens*.

Chapter 15

A PROLONGED
BRAIN DEVELOPMENT

As already mentioned in chapter 13, that humans and the apes are closely related and very similar was evident during the nineteenth century. In *The Descent of Man*, Darwin made the following remark:

> Monkeys are born in almost as helpless a condition as our own infants; and in certain genera the young differ fully as much in appearance from the adults, as do our children from their full-grown parents. It has been urged by some writers as an important distinction, that with man the young arrive at maturity at a much later age than with any other animal: but if we look to the races of mankind which inhabit tropical countries the difference is not great, for the orang is believed not to be adult till the age of from ten to fifteen years. Man differs from woman in size, bodily strength, hairyness, &c., as well as in mind, in the same manner as do the two sexes of many mammals. It is, in short, scarcely possible to exaggerate the close correspondence in general structure, in the minute structure of the tissues, in chemical composition and in constitution, between man and the higher animals, especially the anthropomorphous apes.[1]

Earlier, Thomas Henry Huxley had written in his book *Man's Place in Nature*:

> So that it is only quite in the later stages of development that the young human being presents marked differences from the young ape, while the latter departs as much from the dog in its development as the man does. Startling as the last assertion may appear to be, it is demonstrably

279

true, and it alone appears to me sufficient to place beyond all doubt the structural unity of man with the rest of the animal world, and more particularly and closely with the apes.[2]

In these excerpts both Darwin and Huxley are eager to show the closeness of humans and apes, which is evident in the figure included in Huxley's book (see figure 15.1). They thus put emphasis on that, even though they both mentioned differences in the development of humans and apes. Could this identification of developmental differences between humans and apes be important?

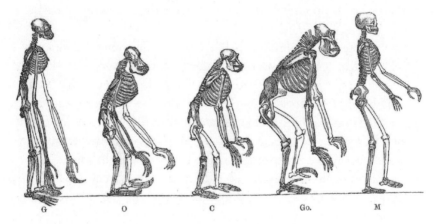

Figure 15.1. The comparison of skeletons of gibbons (G; drawn at larger size), orangutans (O), chimpanzees (C), gorillas (Go), and humans (M), as drawn in Huxley's 1863 book *Evidence as to Man's Place in Nature*.

Well, at that time, naturalists had realized that there is some connection and correspondence between evolution and development, sometimes under the simplistic view that development recapitulates evolution, that is, that the stages that individuals of a species go through during their development are a summary of the stages through which their species evolved to its current state. But nowadays scientists are aware that there is more to it than that. Actually, a major problem behind the public misunderstanding of evolution is the neglect of developmental processes. When we are talking about the evolution of forms, we forget that it is not an adult form that evolves into another one (e.g., avian dinosaurs into birds or our common ancestor to either ourselves or the chimpanzees). What actually evolves is the developmental process that produces the one or the other form. In some cases, this happens because of changes at the genetic level, when genes change, for instance, in terms of their DNA sequence. However, it is also possible for developmental processes to evolve not because of changes in genes themselves but because of changes in when these genes are expressed, or not expressed, during development. Minor changes like these can have a large impact on phenotypes and their evolution.[3]

The respective field of evolutionary developmental biology, or "evo-devo," has provided novel insights about the evolution of organisms. For example, whales and dolphins are mammals but do not have hind limbs. This is the case because even though the development of hind limbs begins in their embryos, it is nevertheless not maintained because they lack a regulator sequence, *Hand2*, which in turn regulates the production of Sonic hedgehog (Shh), a protein with a key role in the regulation of initial limb development. It seems that reduction of Shh expression may have led to the loss of limbs. Therefore, the absence of limbs in whales and dolphins is due to changes in regulatory sequences that affect development.[4] Another example are bat wings, which are in fact not their fore-limbs but their digits. During embryonic development, the digits in bats are initially similar in size to those of mice, but later they lengthen enormously. It seems that bone morphogenetic protein 2 (Bmp2) causes the lengthening of digits in the embryonic forelimbs of bats as its expression is increased therein compared to mouse or bat hind-limb digits. This means that changes in regulation and increased expression of the Bmp2 protein is a major factor in the developmental, and probably in the evolutionary, elongation of bat forelimb digits.[5] Researchers have found several other cases in which large morphological changes can occur by minor changes in underlying molecular networks and mechanisms.

A landmark paper published in 1975 emphasized the astonishing similarities between humans and chimpanzees at the molecular level and noted that "their macromolecules are so alike that regulatory mutations may account for their biological differences."[6] Humans and chimpanzees were two species that were extensively studied at the time, both at the biochemical and at the morphological level. The results of the different biochemical methods used until then (such as amino acid sequencing, that is, finding the sequence of the amino acids—the building blocks— of a protein) produced results supporting the conclusion that the differences between the two species at the protein level were few and therefore not enough to account for their differences at the level of the organism. As the "average" human protein was more than 99 percent identical to the respective protein in chimpanzees, the respective differences in the genes themselves did not seem to be enough to account for the major differences between humans and chimpanzees at the level of the organism. Therefore, these differences could be due to a relatively small number of genetic changes in DNA sequences affecting the expression of their genes.

Indeed, earlier DNA hybridization studies and later DNA sequencing studies found the differences in the DNA sequence of humans and chimpanzees to be about 1 to 2 percent (as mentioned in chapter 13, a more recent estimate of the differences between the human and the chimpanzee genomes is 4 percent).[7] However, even though the DNA sequences of humans and chimpanzees are very similar, only 20 percent of proteins are identical between the two species. This means that 80 percent of their proteins are different, but even this may not be enough to explain the differences observed at the level of the organism. Therefore, regulatory genes should also have a role.[8] Before these DNA and protein comparisons were made, Steven Jay Gould wrote a book-length account of how changes in development can bring about evolutionary change. He focused on "changes in the relative time of appearance and rate of development of characters already present in ancestors," a phenomenon called heterochrony.[9] Gould noted that "heterochronic changes are regulatory effects—

they represent a change in rate for features already present. ('New' features may arise either from changes in structural genes or from shifts in regulation yielding complex morphological effects that we choose to designate as novel)."[10] This means that heterochrony is about when something that already exists appears, and at what rate.

Here is a fictitious example of how this may happen at the molecular level and at the level of the organism. Imagine an ancestral species of beetles (G) that have a gray color. Imagine that this is due to the expression of particular genes that produce a certain amount of a molecule (M) in the exoskeleton that makes it look gray. Now imagine three descendant species of beetles, in which development lasts about the same time. One of them (D) has retained the developmental processes of the ancestral species. However, in the two others there has been a change in the expression of the genes related to the color of the exoskeleton. In species W the expression of the genes related to color has now been limited to a shorter period of time. As a result, they do not have much M in their bodies, and therefore the individuals of this species are white. In contrast, in species B the expression of the genes related to color has been expanded significantly during developmental time, and as a result its individuals have a large amount of M in their bodies and they are thus black (see figure 15.2). This example shows how phenotypic changes such as the change in the color can take place not because of changes in the genes themselves but because of changes in the timing of their expression during development. Actual examples of this kind are well known, including the relative timing of egg hatching and segment development in arthropods, or the expression of opsin genes in fish.[11]

In other cases it is the duration of development that can change. The development of the body can be delayed in relation to the maturation of the reproductive organs, a phenomenon called neoteny. The outcome of this is the retention of ancestral juvenile characteristics by later developmental stages of the descendants. This phenomenon is called paedomorphosis, and it is a case of heterochrony (but it should be noted that neoteny is not the only cause of paedomorphosis). It has long been

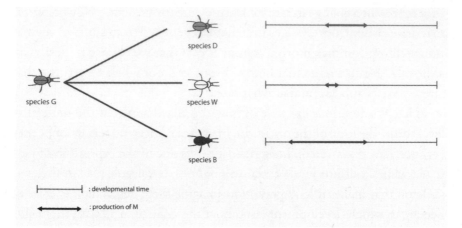

Figure 15.2. A fictitious example of heterochrony in development. In ancestral species G the temporal expression of genes related to exoskeleton color was the same with that of contemporary species D. However, compared to G, this expression has been shortened in species W and extended in species B. As a result, less of the molecule M related to color is produced in the individuals of species W, which are white, and more M is produced in the individuals of species B, which are black.

pointed out that the skulls of juvenile chimpanzees and adult humans are quite similar; however, during chimpanzee development this similarity disappears. The skulls of an infant chimpanzee and a human baby, as well as those of an adolescent chimpanzee and an adult human are quite similar in bone structure and the overall shape. However, the skull of the adult chimpanzee is further differentiated, and it ends up having a smaller cranial capacity and a larger jaw. In other words, it is as if chimpanzees and humans have a similar skull development until a certain point. Humans stop there, whereas chimpanzees continue to differentiate further. This process can be considered in the opposite way. If we imagine that there are three different stages in the development of the chimpanzee skull (1. infant; 2. adolescent; 3. adult), it seems that human development is delayed so that humans manage to reach stage 2 but not stage 3. In this way, what is a juvenile skull in chimpanzee becomes the adult skull in humans. Neoteny is, in other words, the extension of the early stages of development, with later stages being shortened or eliminated.[12]

As a result of this heterochronic change, humans remain at a developmental stage in which they have a skull that is relatively larger than that of chimpanzees. Indeed, this is one of the most striking features of human evolution. It seems that for several millions of years there was no significant difference in the mean cranial volume of our ancestors. For instance, *Australopithecus afarensis*, who lived approximately between 3.89 and 2.9 million years ago, had a mean cranial volume of 433 cm^3 (range: 385–550 cm^3); *Australopithecus africanus*, who lived approximately between 4.02 and 1.9 million years ago, had a mean cranial volume of 454 cm^3 (range: 391–568 cm^3). This means that for approximately 2 million years, the cranial volume of *Australopithecus* did not change significantly. In contrast, the size of *Homo* seems to have increased significantly during approximately the last 2 million years. In the beginning of our lineage, we find the Dmanisi hominins, who lived between 1.85 and 1.77 million years ago and who had a mean cranial volume of 645 cm^3 (range: 546–790 cm^3). Then, in *Homo ergaster*, living approximately between 2.27 million and 870,000 years ago, we see a mean cranial volume of 796 cm^3 (range: 715–909 cm^3). Things change further in *Homo erectus*, who lived between 1.85 million years ago and 27,000 years ago, where we see a mean cranial volume of 981 cm^3 (range: 656–1,300 cm^3). This indicates a significant increase of cranial volume. *Homo heidelbergensis*, who lived between 700,000 and 100,000 years ago, had a mean cranial volume of 1,158 cm^3 (range: 1,057–1,250 cm^3), pretty close to but still a bit lower than our own. With *Homo neanderthalensis*, living between 130,000 and 40,000 years ago, we arrive at an average cranial volume of 1,420 cm^3 (range: 1,172–1,740 cm^3), which is very close to our own average of 1,457 cm^3 (range: 1,090–1,775 cm^3). However, at the same time we should not forget *Homo floresiensis*, who lived between only about 100,000 and 60,000 years ago and had a mean cranial volume of 417 cm^3.[13]

Therefore, the rough picture is that after a rather static phase of at least 2 million years, between the split from chimpanzees and the emergence of the genus *Homo*, during which there was no significant increase in cranial volume, we see a striking increase during the next 2 million years until today. During that time, we can roughly say that the cranial volume doubled during the first million years and doubled again during the second million years until our time. So, what conclusion can be made from this evidence? Relative to their body size, the adult forms of our ancestors had a smaller skull than we do. However, their infants initially had a relatively larger skull for their body size, and during development the ratio of skull size to body size became smaller. This is the case for the chimpanzees and ourselves, too. Something happened to our ancestors after the split from chimpanzees that made them retain a larger ratio of skull size to body size than before. This provided them with larger cranial volumes. But such growth did not stop there. For some reason, the ratio continued to increase up to our species. Larger skulls mean larger brains; but what does it entail? Is that what could make this change a critical one that had an impact on our evolution, and perhaps a turning point?

To answer this question, it is necessary to consider a different kind of evidence. A study analyzed the levels of gene expression in the brains (in particular in the prefrontal cortex, the front part of the brain related to the control of actions) of 39 humans, 14 chimpanzees, and 9 rhesus macaques during development after birth. The researchers detected and studied the expression of 7,958 genes in these individuals, and they found that gene-expression levels changed with age, as well as that 71 percent of the nearly eight thousand genes expressed in the human brain changed significantly during development after birth. Researchers also found that in humans and chimpanzees, gene expression changes during brain development occur early in development, with half of these changes occurring during the first year of life. However, the changes in the timing of expression of individual genes differed between humans and chimpanzees. To further investigate whether these expression differences between humans and chimpanzees reflected changes in the timing of developmental changes (heterochrony), the researchers developed a test that could also distinguish whether the heterochronic expression changes were in the form of delays (neoteny) or accelerations. This test was applied to data from 14 humans and 14 chimpanzees closely matched with respect to age and sex, aiming at assigning the changes in gene expression in one of the following categories: (a) human neoteny–expression changes, with human expression corresponding to that in younger chimpanzees; (b) human acceleration–expression changes, with human expression corresponding to that in older chimpanzees; (c) chimpanzee neoteny–expression changes, with chimpanzee expression corresponding to that in younger humans; (d) chimpanzee acceleration–expression changes, with chimpanzee expression corresponding to that in older humans. From the 3,075 genes expressed, 299 could be assigned to one of categories a–d, with 38 percent being human neotenic genes (category a), and the remaining 62 percent being assigned in the three other categories. This significant excess of genes showing neotenic expression in humans supports the role of neoteny in human evolution, and suggests that differences in the developmental timing of gene expression between humans and chimpanzees may reflect differences in sexual and cognitive maturation between these species.[14]

A more recent study examined the changes in gene expression that take place during brain development after birth in the prefrontal cortex and the cerebellum of 23 humans, 12 chimpanzees, and 26 rhesus macaques. Attention was paid to those changes in genes that are specific to humans, that is, that the changes in gene expression during development differed significantly between humans and chimpanzees, and between humans and macaques, but not between chimpanzees and macaques. The researchers found 702 genes that had human-specific expression and 55 genes that had chimpanzee-specific expression in the prefrontal cortex, as well as 260 genes with human-specific expression and 82 genes with chimpanzee-specific expression in the cerebellum. This excess of human-specific developmental differences, compared to chimpanzee-specific ones, observed in the prefrontal cortex and the cerebellum is important, as both of these brain regions are considered to be related to complex and possibly human-specific cognitive functions such as social behavior, abstract thinking, reasoning, and language. A striking difference observed for a particular set of genes was that gene expression in the prefrontal cortex increased after birth and started decreasing five years after birth in humans, whereas in both chimpanzees and macaques gene expression started decreasing soon after birth. Another important finding was that these genes are associated with synaptic functions, that is, the functions related to the connections (synapses) between neural cells, and the transmission of signals through them. It can therefore be the case that the extension of gene expression related to synapses in the prefrontal cortex, which is related to the extension of brain development and wiring and which seems to have occurred after the separation of the human and the chimpanzee lineages, may be associated with the emergence of human-specific cognitive characteristics and intelligence.[15]

Complementary evidence for this conclusion comes from a study that attempted to estimate the genetic bases of brain size and the organization of brain cortex in chimpanzees and humans by studying phenotypic similarities between individuals related to one another. If the human brain is more plastic and can therefore be more extensively modified by environmental factors than the chimpanzee brain can be, then there should be differences in the heritability of brain size and cortical organization in chimpanzees and humans. It must be noted that heritability is the amount of phenotypic variation that is due to variation in genes, and not a measure of how genetic or how environmental a phenotype is (in other words, heritability is not a measure of how *inheritable* a characteristic is). This simply means that if parents and offspring exhibit a variation in a characteristic, we can estimate how much of this variation is due to variation in their genes and how much is due to variation in their environment. In other words, we can explain differences in something on the basis of differences in something else. For this purpose, researchers analyzed MRI scans of 206 chimpanzees and 218 humans. The results showed that the heritability for brain size was very high in humans and quite high in chimpanzees—but, at the same time, the brain size heritability of chimpanzees was quite lower than it was for humans. However, it was found that cerebral cortical anatomy in humans was substantially less genetically heritable than it was in chimpanzees. This indicates a relaxed genetic control and a greater plasticity in the morphology of the human cerebral cortex, a property that makes our brain more responsive to environmental influences during its development. In other words, whereas the size of our brain may not differ that much from that of our parents, the wiring among its cells can be different, reflecting our experiences during development.[16]

To put things simply, this evidence supports the conclusion that the human brain continues to grow and develop (in the sense of forming connections among brain cells) long after our birth. Whereas chimpanzees, our closest relatives, reach sexual maturity around the age of 8–9 years, we get there at the age of 13–14 years. Now, if during the first five years of our life our brain develops in some important aspects, something not happening in chimpanzees, then it would seem as if evolution has added a crucial five-year period between infancy and pre-adolescence that is unique to us. This is the period that we call childhood, during which we learn a lot about the world. It seems to be the case that the same neuronal networks that are largely fixed in chimpanzees at birth remain plastic and keep developing in us. Thus, during our childhood our brain cells keep making connections with one another, depending on our experiences from the world, and learning behaviors, languages, and more. In this sense, we have an extended period of synaptic plasticity, or flexibility in making connections among brain cells, during which our brains are remodeled. This increased level of synaptic plasticity may be related to our increased learning abilities.

Indeed, a study of cognition about causal, object, spatial, and temporal relations in capuchins, macaques, chimpanzees, bonobos, and humans has shown some cognitive features common among all these five species, as well as a heterochronic variation of these features during development. For example, by the age of five years, members of all of these five species are able to classify objects on the basis of their properties, such as their form or the material of which they are made. However, there are also important differences. When capuchins and macaques are given square columns and cylinders, they can separate one category by putting together cylinders only. Bonobos and chimpanzees can progress to classifying two categories but not three. In contrast, human development progresses from one category to two categories to multiple categories and even to hierarchic-category classification. At the same time, the ability for classification is evident in very young ages in humans, develops very fast, and stops developing at a relatively old age; the ability for classification is evident in intermediate ages in bonobos and chimpanzees, develops in intermediate pace, and stops developing at an intermediate age; finally, the ability for classification is evident in older ages in capuchins and macaques, develops in a relatively slow pace, and stops developing at a relatively young age. These findings, which show that similar abilities differ in the timing of their development, are consistent with the idea of heterochrony in cognitive development.[17]

To summarize, neoteny seems to be an important factor in human evolution because the prolongation of development and the retention of fetal growth rate may have caused the increase in brain size that is characteristic of our evolution. This has clearly been very important for our evolution, as our cognitive abilities allow several skills that no other species exhibits. However, the prolonged development of our brain that has resulted in our relatively increased brain and skull size also has disadvantages. One disadvantage is that whereas the weight of the human brain is approximately 2 percent of the weight of the human body, it consumes approximately 20–25 percent of its energy. Thus, a human brain requires approximately 280 to 420 calories per day, whereas that of a chimpanzee about 100 to 120. This energy requirement may not seem that much today, now that many of us have food at our disposal, but it would be a lot for a hunter-gatherer who would need to consume a lot more food than a chimpanzee in order to sustain his brain's functions. Another disadvantage relates to birth. The head of a human newborn is about 125 millimeters long and 100 millimeters wide, whereas the birth canal can about 113 millimeters long and 122 millimeters wide. This entails that for a newborn to be able to pass through its mother's birth canal, it must enter her pelvis facing sideways and then make a 90-degree turn. This is why birth is slower and much more difficult in humans than in most other primates.[18] Clearly, there are costs for having large brains.

Now the question that begs an answer is this: when did the prolongation of gene expression related to synapses among neural cells, and therefore the prolongation of our neural plasticity, occur? To answer this question, we need to look at fossils. On one hand, the patterns of cranial growth of *Homo neanderthalensis* and ourselves seem to be similar, exhibiting a rapid growth after birth not found in *Homo erectus* or chimpanzees. On the other hand, the rate of dental maturation in *Homo neanderthalensis* seems to have been rapid and similar to that of *Homo erectus* and chimpanzees, and different from the delayed maturation observed in fossils of modern humans. This means that certain aspects of development were shared between the ancestors of modern humans and *Homo neanderthalensis*, whereas other changes in developmental patterns may have occurred after the separation of the two species. Indeed, there is evidence that human brain development changed fundamentally shortly after this separation and before the emergence of modern humans. For example, a comparison between the modern human genome and the genome of Neanderthals has shown that *MEF2A*, one of the potential transcriptional regulators of extended synaptic development in the human cerebral cortex, has been an object of natural selection in humans during the last 500,000 years. Particular changes in the DNA sequence of this regulator, for example, a single change of one base to another leading to a regulatory change, could have rapidly increased in frequency in a population, probably because it provided an advantage in survival and reproduction to its bearers.[19]

Molecular changes like this one that could have resulted in changes in the timing of developmental events and eventually in the size and features of our brains are **turning points**. These changes are unpredictable and therefore contingent *per se*. As already explained in part 2 of the present book, one cannot tell in advance when a change in the sequence of DNA will take place. However, once this happened it could have had a major impact, and, therefore, the subsequent evolution of our species has been contingent *upon it*. Molecular changes in regulatory sequences could have caused a prolonged development and an extended period of brain growth relative to chimpanzees, which in turn caused us to have relatively larger and more plastic brains, and eventually a relatively higher intelligence. This prolonged development itself was also a **turning point**, because it was contingent *per se* and our subsequent evolution was contingent upon it. As Gould described it:

> The evolution of consciousness can scarcely be matched as a momentous event in the history of life; yet I doubt that its efficient cause required much more than a heterochronic extension of fetal growth rates and patterns of cell proliferation. There may be nothing new under the sun, but permutation of the old within complex systems can do wonders.[20]

Chapter 16

OUR BIOCULTURAL EVOLUTION

Even though we may think that our species, *Homo sapiens*, now dominates Earth (because we often forget about how successful other organisms, such as bacteria, are), we should not overlook the fact that there have been several other human species on this planet. Although the evidence is fragmented and there is a lot of ongoing research on that, we are aware of several other human, or human-like, species, which have lived on Earth, among them *Homo ergaster*, *Homo erectus*, *Homo heidelbergensis*, *Homo neanderthalensis*, *Homo habilis*, *Homo rudolfensis*, and *Homo floresiensis*. What happened to all of these other species of the genus *Homo*? Modern humans first appeared in Africa about two hundred thousand years ago, and then spread across the world—but note that even as I am writing these lines, new evidence has emerged that this might had happened approximately one hundred thousand years earlier than that.[1] Some of these species preceded our own, whereas some others seem to have coexisted and interacted with our ancestors. The current picture as it emerges from the fossils that have been found is rather fragmented; however, it indicates the considerable diversity in our lineage during the last five hundred thousand years (see figure 16.1). We do not know the exact relations among the various species, but we at least have an idea of when they lived on Earth. As is obvious, our ancestors coexisted for some time with the Neanderthals (*Homo neanderthalensis)*, and the so-called Denisovans, another extinct species or subspecies of *Homo*. Let us briefly see what we currently know about these two close relatives of ours.

Fossils of Neanderthals have been found in various places in Europe, Asia, and the Near East. Some of their distinctive characteristics are a

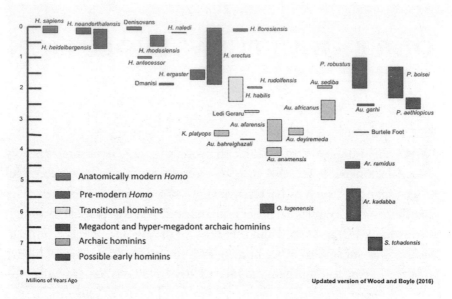

Millions of Years Ago

Updated version of Wood and Boyle (2016)

Figure 16.1. The hominin fossil record across geological time, based on a conservative estimate of the temporal ranges of the hominin species, site collections, and individual fossils referred to in Bernard Wood and Eve K. Boyle, "Hominin Taxic Diversity: Fact or Fantasy?" *American Journal of Physical Anthropology* 159, no. S61 (2016): 37–78. The bottoms and tops of the continuous columns represent, respectively, the currently available dates of each species' first and last appearance (the figure above is not reprinted from the aforementioned article but is an updated version from June, 20 2017, kindly provided by Bernard Wood and Eve K. Boyle, and it is reprinted here with their permission) See also Bernard Wood, "Evolution: Origin (s) of Modern Humans," *Current Biology* 27, no. 15 (2017): R767–69.

forward-jutting face with a big nose; thick, protruding brow ridges; and a projected occipital bone at the back of the skull. As was already mentioned in the previous chapter, the average cranial volume of the Neanderthals was 1420 cm³, which is quite large and close to our own. Several Neanderthal fossils have been found at various places in the world, but what is more striking is that researchers have been able to reproduce the genome sequence of the Neanderthals. In contrast, we do not know much about the Denisovans, as their exact relation to the Neanderthals and our ancestors is unclear. A main reason for this is that we have found only a few bones. However, in this case too, it has been possible to study parts of their DNA. Let us now see in some detail what the study of DNA of the Neanderthals and the Denisovans tells us about our recent past (that is, a few thousand years in geological time, so quite recently).

The first Neanderthal fossil bones were found as early as 1856 in a cave of the Neander valley in Germany. Whereas retrieving DNA sequences from specimens older than one hundred thousand years is difficult, if not impossible, because DNA is usually damaged, during the 1990s it was possible to extract DNA from the Neanderthal specimens and analyze it. In particular, researchers were able to find the sequence of a particular region of mitochondrial DNA. Mitochondrial DNA is a small DNA molecule found within the mitochondria, the organelles that produce energy in our cells. These, as well as their DNA, are replicated and are passed on to the cells emerging from cell division. The special feature of this DNA is that it is passed on only from a mother to her offspring. The reason for this is that during fertilization it is only the nucleus of the spermatozoon that enters the ovum, while its mitochondria mostly stay out of it. As a result, the emerging embryo will have the mitochondria of the ovum, and therefore the mitochondrial DNA (mtDNA) that comes from the mother. For this reason, mitochondrial DNA can be very useful to trace ancestry along a matrilineal lineage. Exactly because it is passed on from mother to children, it can form the basis for finding ancestry from children to their mother, to their maternal grandmother, to their maternal great-grandmother, and so on (the respective molecule for doing the same thing along a patrilineal lineage is the DNA of the Y chromosome). The analysis of the Neanderthal mtDNA supported the conclusion that it indeed came from the fossil and not from some other source (e.g., the humans that found the fossil) and that it was very different from human mitochondrial DNA. The conclusion that the researchers reached was that there was no contribution of Neanderthal mtDNA to modern humans.[2]

In 1997 and 2000, new excavations took place at the Neander valley, and they brought to light more fragments of human bones and artifacts of the Paleolithic era. These more recent excavations resulted in the discovery of sixty-two human skeletal fragments belonging to at least three individuals—two adults (one of whom was the one discovered back in 1856) and a youth. The dating of these specimens showed that these individuals lived about forty thousand years ago. But, most interestingly, researchers were able to extract DNA from the bones of one of these individuals and analyze it. The DNA analysis resulted in the determination of a new mtDNA sequence that had three differences from the DNA sequence of the initial Neanderthal mentioned above, suggesting that it came from an individual not only different from but also maternally unrelated to the first one.[3] Eventually, in 2010, a draft sequence of the Neanderthal genome from three individuals was completed. This was compared to the genomes of five modern humans from different parts of the world, and several differences between the Neanderthal genome and the human genomes were identified, including some differences in genes involved with metabolism, cognitive abilities, and cranial morphology. It was concluded that the Neanderthal genome was on average more similar to the genome of present-day humans from Eurasia than it was to those from Africa. At the same time, present-day humans from Eurasia have regions in their genome that are more similar to those in Neanderthals and more distant from other present-day humans. The conclusion made from the comparison was that between 1 and 4 percent of the genomes of people in Eurasia were derived from Neanderthals. Whereas this amount is small, it indicates that some kind of interbreeding took place between the Neanderthals and the ancestors of modern humans. This does not change the fact that most of the current human genetic variation is of African origin, but it is important to know.[4]

At around that time, another discovery took place. A finger bone from a hominin was found in Denisova Cave in the Altai Mountains in Russia. That was part of the skeleton of an unknown type of hominin. The analysis of the mtDNA obtained from that bone supported the conclusion that that individual had a common ancestor with modern humans and Neanderthals, and that that ancestor lived about 1 million years ago. The Denisovan mtDNA had almost twice as many differences from the mtDNA of present-day humans when compared to the Neanderthal mtDNA. This indicated that the common ancestor of all three groups should be older than the common ancestor of present-day humans and the Neanderthals, and that the separation of the latter took place after the separation from the Denisovans. It was also concluded from the stratigraphy, that is, the order and relative position of rock layers in the cave, that the Denisovan individual lived between thirty thousand and fifty thousand years ago, and so coexisted with both present-day humans and the Neanderthals.[5] A subsequent study sequenced the Denisovan genome of the nucleus in order to further investigate its relationships to Neanderthals and present-day humans. What was found was that the average difference of the Denisovan nuclear genome (i.e., the DNA found in the nuclei of our cells that is most of our DNA) from present-day humans is similar to that of Neanderthals. Further analysis supported the conclusion that the Denisovan individual and the Neanderthals are both descended from a common ancestral population that was already separated from the ancestors of present-day humans. In other words, the Denisovans and the Neanderthals were closer to each other than they were to ourselves.[6]

A more recent study that performed an analysis of genomes of a Neanderthal, a Denisovan, and present-day humans has shown that there was interbreeding among the Neanderthals, the Denisovans, and early modern humans. The available evidence supports the conclusion that these groups, as well as a fourth group of an unknown archaic hominin, met and had offspring on various occasions. Even though the extent of this interbreeding seems to have been generally low, it indicates the transfer of genes: (a) from the Neanderthals to the ancestors of many present-day human groups in Asia; (b) from the Neanderthals to the Denisovans; and (c) from an unknown archaic hominin to the Denisovans (see figure 16.2).[7] There is also evidence that genes may have been transferred from the Denisovans to human groups in Asia, as well as that early modern humans mixed with Neanderthals when they arrived in Europe from Africa. This could have provided our ancestors with the ability to adapt to the Eurasian environments, because of the introduction of potentially advantageous alleles from the other groups, which might have never appeared due to mutations alone. The Neanderthals and the Denisovans had lived in Europe and Asia for hundreds of thousands of years, and so they might have had alleles that were favorable for the conditions in which they lived. Thus, when our ancestors arrived to Eurasia from Africa and interbred with these groups, they further produced offspring that might have acquired these favorable alleles that could have then become prevalent in the human population through natural selection exactly because they were advantageous.[8]

The evidence presented so far supports the conclusion that we have not been the only human species on Earth even in (relatively) recent times, as other humans have coexisted and interacted with our ancestors about 30,000–40,000 years ago. This is important to note before we get to an explanation of why we are currently the only human species on Earth later in this chapter. It is important to note that the available evidence, an overview of which is given in figure 16.1, shows how complex human evolution has been, which is completely inconsistent with any idea that humans may have been intelligently designed. As I show later in this chapter, our uniqueness and special status in this world seems rather to be the outcome of particular contingencies.

You may be wondering at this point why the tree in figure 16.2 is complicated and not as clear as the one in figure 13.1, which showed orangutans, gorillas, chimpanzees, and humans. There are three kinds of problems to take into consideration. The first is that extracting DNA from fossils and finding DNA sequences as above is neither simple nor always effective. There are two main obstacles in doing this. The first is that DNA may not exist at all in fossils, and when it does it is more often than not so degraded that only small DNA fragments can be obtained.

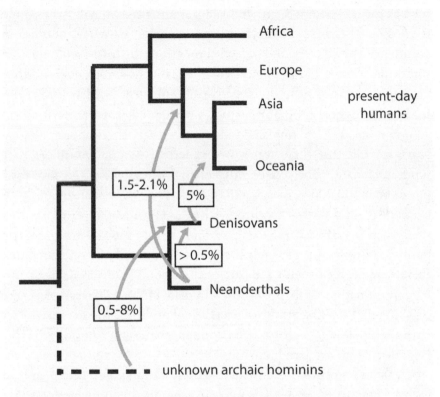

Figure 16.2. An overview of the interbreeding among our ancestors, the Neanderthals, the Denisovans, and a potentially existing but unknown archaic hominin group. The Denisovans and the Neanderthals are most closely related, we share a common ancestor with them, and all three groups should have a common ancestor with the potentially existing archaic hominin species. The gray arrows indicate the transfer of genes from one group to another (adapted from Svante Pääbo, "The Human Condition–a Molecular Approach," *Cell* 157, no. 1 [2014]: 216–26.)

Putting together separate DNA fragments and finding their actual order in the genome is far from easy, and it requires diligent work in order to be achieved. Another difficulty with DNA extraction is that DNA in fossils can be significantly contaminated by DNA from other sources, such as bacteria, animals, and most importantly the humans who found the fossils. Therefore, before any conclusion can be made, one has to ensure that the DNA found indeed comes from the fossils and not from some other source. For example, two articles that described nuclear DNA sequences obtained from the same Neanderthal fossil had arrived at different conclusions. One of the studies concluded that the most recent common ancestor of humans and Neanderthals lived about 706,000 years ago and that there was no Neanderthal genetic contribution to modern humans.[9] The other study concluded that the divergence took place about 500,000 years ago and that there was some transfer of DNA between them.[10] As the two studies had used different methods, other researchers re-analyzed the data, assuming that at least one of these two published studies was probably incorrect. The re-analysis concluded that there were problems with the data of the second study, most likely due to contamination with present-day human DNA.[11] Nevertheless, the conclusion of the first study eventually had problems too because, as we have already shown, there has been a minor genetic contribution from the Neanderthals to modern humans.

A second problem in constructing these "family trees" is that, at least for the time being, only a few ancient genomes are available for study. Such a very small sample cannot be considered a representative of the respective population, because specific individuals do not necessarily carry the alleles that were prevalent in their population. Clearly, we need representative samples; these will probably never be as large as those that are available for present-day humans, of course, but hopefully new findings will allow for a more detailed view of these ancient humans' genomes. Another important point is that even if we are able to identify that Neanderthals and Denisovans had this or that "human" gene, we cannot really know what its role was. We already know that the function of individual genes in present-day humans depends on which other genes are found in one's genome (as discussed in chapter 6). Therefore, one can imagine that it is even harder to know what a gene may have done in a living Neanderthal or Denisovan. At this point, there are significant limitations in the inferences that are possible from the study of ancient genomes.[12] We should also not forget that what we have been referring to as Denisovans in the present book is actually not a population, not even an individual, but just a few bones.

A third kind of problem is that the comparisons made in figure 16.2 are at the level, or close to the level, of species. This is microevolution, not macroevolution. What is being compared is not reproductively isolated, distinct species such as those depicted in figure 13.1. Rather, what we could have at best are populations of humans that were moving around, interacting with each other and, most importantly, interbreeding. Therefore, it is not possible to clearly distinguish between them. This is why we have no idea whether or not the Denisovans are actually a distinct species. This is also why experts disagree about whether the Neanderthals are indeed a different species than our own (*Homo neanderthalensis*, whereas we are *Homo sapiens*) or actually a subspecies of our own species (*Homo sapiens neanderthalensis*; in this case, we should be *Homo sapiens sapiens*). To this we should of course add the consideration that classification of organisms into species is a cultural construction and therefore relatively arbitrary, rather than inherently natural.[13]

Now, the big question is this: why are we the lone survivors among the hominin groups, and what caused the extinction of the others, like the Neanderthals and the Denisovans? For the Denisovans, there is not much to say, as we do not really know if they were a distinct population. For the Neanderthals, it has been commonly assumed that they disappeared in the struggle for existence with our ancestors. Several different hypotheses have been proposed, suggesting that, compared to our ancestors, the Neanderthals: did not have their "complex symbolic communication systems" and "fully syntactic language"; had limited capacity for innovation; were less efficient hunters; had weaponry that was of inferior efficacy; had a narrower diet; did not use traps and snares to capture animals; had smaller social networks; had a smaller population; had a simpler hafting technology; were less tolerant of the cold climate around forty thousand years ago; and/or were indirectly affected by the eruption of the Mount Toba volcano at seventy-five thousand years ago. Given the currently available archeological data, none of these hypotheses seems to be supported. The replacement and rapid extinction of Neanderthals could be explained if there were significant cognitive, technological, and demographic differences between themselves and our ancestors. However, their archaeological record is not different enough from our ancestors' to support the idea that they were somehow technologically and cognitively inferior to them, particularly if you also consider that their brains were about the same size as ours. A more plausible explanation is that extensive interbreeding probably led to their assimilation from our ancestors and the disappearance of the characteristic Neanderthal morphology from the fossil record.[14]

But, then, why was it us who assimilated them, and not the opposite? A plausible explanation is that it was our distinct human culture that made the difference.[15] The term *culture* in the vernacular sense is used to refer to the beliefs, values, and traditions of particular populations. However, researchers who pioneered the study of the connection of culture to human evolution have suggested the following definition: "Culture is information capable of affecting individuals' behavior that they acquire from other members of their species through teaching, imitation and other forms of social transmission."[16] Under this definition, one can imagine cultural change as a form of evolutionary change whereby advantageous culturally transmitted information, such as knowledge, beliefs, values, and skills, are passed on to descendants. As I explain below, these can have a major influence on biological evolution. But before this, let us consider an example of cultural evolution.

Linguistic evolution is a good example of cultural evolution. It seems that languages evolve largely independently from one another, even though the phenomenon of borrowing and using words—especially between neighboring populations—does happen. Languages generally exhibit variation in time and in space, and the latter is a lot easier to study, as it can be done in contemporary populations. There can be phonological, semantic, and grammatical variation in languages.[17] Phonological variation has to do with the pronunciation of words and the differences between, for instance, the English of a British and an American, or the French of a French and a Canadian. An example of grammatical variation has to do with the order of subject, verb, and object in a sentence. This is the order one finds in English, French, and Greek, but not necessarily all the time. For instance, in French we may say, "*je t'envoie le cadeau*," and in Greek we may say, "*εγώ σου στέλνω το δώρο*"; however, in English we would never say, "I you send the gift" (which is the equivalent for the phrases in French and Greek just mentioned), but rather, "I send you the gift" or "I send the gift to you." There can also be semantic variation. For instance, in both English and French the word "table" refers to the same piece of furniture (with a flat top and one or more legs), even though it is pronounced differently in the two languages. Furthermore, in English the word "table" is also used to refer to a scheme for systematically displaying a set of facts or figures in columns, whereas the respective word in French is "*tableau*." A more striking example can be found in the Greek-rooted English words "empathy" and "sympathy." In English, "empathy" is the ability to understand and share the feelings of someone else, whereas "sympathy" is the feeling of sorrow for something that happened to someone else. In contrast, in modern Greek, "empathy" is a synonym for hatred or intense dislike for someone, whereas "sympathy" is used to express a feeling of liking or fondness. All of these are examples of how languages with a common origin have evolved across time, as words are being adapted to various uses.

In a similar sense, cultural norms and practices that contribute to survival and reproduction can evolve over time. An initially simple object used for opening seeds, such as a stone, or for hunting animals, such as a sharp branch, can evolve to more sophisticated forms. This evolution is based on various trial-and-error attempts made by different individuals of the same population who end up having a collective knowledge about how to made a particular tool or how to overcome difficulties in the environment, such avoiding poisonous foods, building secure shelters, or finding food. This collective knowledge can be passed on to future generations and may further evolve over time as subsequent generations contribute to it. In this sense, particular individuals with particular cultural characteristics could have been more effective in surviving and reproducing, thus passing their genes and their cultural information to their descendants.

It could actually be the case that culture evolves faster than genes, thus creating novel environments that favor the selection of particular alleles over others. In such cases, genes and culture co-evolve. A well-studied case of this kind is lactose intolerance. Most humans cease to have the ability to digest lactose, a sugar found in milk, in childhood. However, in northern European populations as well as in sheep/cattle-breeding populations from Africa and the middle East, the ability to digest lactose persists into adulthood; this is known as lactose tolerance. The differences in the ability to digest lactose relate to genetic variation near the *LCT* gene that produces the enzyme lactase, which breaks down lactose. Researchers have found a strong correlation between the frequency of lactose tolerance and its related alleles and a history of dairy farming and drinking milk. This led to the hypothesis that dairy farming and drinking milk favored those individuals who could digest lactose in adulthood and so the respective alleles gradually increased in frequency in those populations. The available evidence supports this hypothesis rather than the idea that the presence of the lactose-tolerance allele allowed dairying to spread, or that this happened for some other reason that had nothing to do with dairying.[18] Another example of cultural changes that have brought about changes in the genetic structure of human populations is that between agriculture and the number of copies of the gene related to starch digestion. For instance, individuals in populations with a recent history of diets rich in starch, such as agricultural ones, have been found to have more copies of the salivary amylase gene (*AMY1*), the gene encoding the enzyme amylase, which breaks down starch—as well as having higher levels of the enzyme amylase in their saliva—than individuals in populations with traditionally low-starch diets. It seems plausible that higher copy numbers of the *AMY1* gene and therefore higher protein levels improve the digestion of starchy foods. Thus, individuals having higher copy numbers of the *AMY1* gene had an advantage in survival and reproduction in the agricultural environments.[19] These are some early and fascinating examples of how gene-culture co-evolution can occur. Particular practices, such as dairying and agriculture, which proved to be useful for reproduction and survival, could have driven the evolution of populations of our species to a particular direction. Individuals who had alleles that conferred advantages related to these practices, such those allowing for lactose tolerance and more efficient starch digestion, were better than others in surviving, and therefore in reproducing and passing on those alleles to their descendants.

But how did we manage to develop these cultural practices and environments that in turn drove our evolution? In the last fifty thousand years or so we have expanded all over the globe, with both good and bad consequences for ourselves and the planet. Our cognitive abilities certainly have had a role. We have relatively large brains that allow us to think in order to figure out solutions to problems, as well as to use our hands for making, constructing, building, or manufacturing numerous useful objects. Thus, we have managed, and still manage to survive in a wide range of different environments. But this was a collective, rather than an individual, achievement. The reason for this is that in all human societies, even in the most primitive ones, people depend on collective knowledge in order to survive. Therefore, our success seems to be due to our ability to learn from others, which in turn enables us to accumulate useful information across generations and develop tools and practices that would be impossible for any one of us to develop alone.[20] This collective thinking is a feature not found in any of the apes, and it is by far distinctively human. For instance, research on the performance of orangutans, chimpanzees, and human toddlers in cognitive tasks has shown no difference in the subtests on mental activities (space, quantities, causality), except for social learning, in which humans outperformed the chimpanzees and the orangutans.[21] The ability to learn from others helped us produce many important innovations, such as fire, cooking, water containers, plant knowledge, projectile weapons, levers, wheels, screws, and writing. These innovations posed new environments and further guided our evolution. And they still do. Two factors seem to have been crucial for this: the size of a population and the degree of social interconnectedness of its members. Simply put, the larger a population, the more likely it is for some of its members to come up with ways to make useful objects and solve problems. Furthermore, the more socially interconnected its members are, the more likely it is for the knowledge about these objects to spread through the population.[22]

This provides the answer to the question I asked earlier, about why it was us and not the Neanderthals who assimilated the other. As already mentioned, the Neanderthals had brains of the same size of our own (if not bigger). Brain size is a strong indicator of cognitive abilities in primates; in other words, the higher the brain size, the higher the cognitive abilities of the species. Therefore, and also given the archeological record, there is no reason to assume that Neanderthals were less competent than ourselves individually. However, it could be the case that our ancestors had a larger capacity for collective learning, because they lived in larger populations with a high social interconnectedness among members. On one hand, the Neanderthals lived in small, scattered groups in ice-age Europe. On the other hand, our ancestors came from Africa, where they lived in more resource-dense and warmer environments, which may have allowed life in large and socially interconnected groups. Therefore, our ancestors could have had a significantly higher ability to develop and spread collective knowledge. In other words, it seems that our ancestors had larger "collective brains" than the Neanderthals, and this was the key factor that enabled them to invade Europe and assimilate the Neanderthals.[23]

The emergence of our ability for collective thinking and learning from others was contingent *per se*, and perhaps contingent *upon* our bipedalism (which freed our hands to make and use tools) and our large brains (which continue to develop after we are born and assimilate what we learn). Our expansion across this planet has in turn been contingent *upon* this collective thinking and our ability to learn from others. Therefore, the emergence of our "collective brains" seems to be a major **turning point** in our evolution. This is relatively recent, at least compared to the chromosome fusion, bipedalism, and prolonged brain development that evolved in our ancestors. It may still be underway. But these are among the important turning points that made us what we are: humans—or, to put it another way, bipedal, large-brained ex-apes with 46 chromosomes sharing collective knowledge and learning. These turning points of human evolution are presented in figure 16.3. We do not know why exactly these characteristics evolved, but we have a good understanding of how they evolved, and this is important. Even more important is our understanding of the impact of these characteristics on the course of the evolution of our species. Although we cannot tell what could have happened had they not evolved, it is plausible to imagine our alter-egos: knuckle-walking individuals with 48 chromosomes and ape-sized brains without cultural learning; knuckle-walking individuals with 46 chromosomes and ape-sized brains without cultural learning; bipedal individuals with 46 chromosomes and ape-sized brains without cultural learning; or bipedal individuals with 46 chromosomes and large brains without cultural learning. Any of these outcomes could have been possible, but evolution took the course toward our condition. Was it inevitable? I do not think so. It was contingent *upon* particular turning points, such as those discussed in this part of the present book, which in turn were contingent *per se*.

In concluding part 4 of this book, I want to note that we are not exceptional, but just one species on this planet; we were not inevitable, but are the contingent outcome of critical events; however, we are in some way unique in this planet, as its only bipedal, large-brained, collectively learning inhabitants. Because of these properties, we have the ability to develop technologies and culture, as well as the ability to make good use of them—do not take reading this book for granted, because there is so much involved in doing this, both inside and outside your head. These abilities are a triumph of evolution, in some sense, and we should be happy about them and about how we got here. But this does not entail that we have any special place in this world. Again, we are unique but not inevitable; we are a very small part of this world but not anything special; we should be modest, as we are just humans.

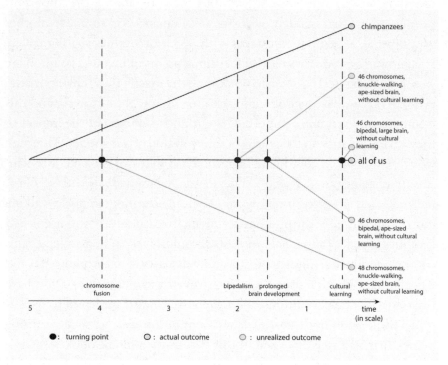

Figure 16.3. Turning points in human evolution. The chromosome fusion, the emergence of bipedalism, the prolongation of our brain development, and our adoption of collective learning have made us who were are.

OVERCOMING THE DESIGN STANCE

T he science fiction film *Jurassic Park*, based on the novel by Michael Crichton and directed by Steven Spielberg, aroused the imagination of spectators in the early 1990s. I was an undergraduate student in biology at that time, excited by the topic of my studies. Therefore, I took that film very seriously when I watched it, and I struggled to understand whether this science fiction scenario could ever become reality, and whether the genetic engineering described in the film was feasible. However, what struck me most in the film was the following dialogue:

> Dr. Ian Malcolm: John, the kind of control you're attempting simply
> is . . . it's not possible. If there is one thing the history of evolution
> has taught us it's that life will not be contained. Life breaks free, it
> expands to new territories and crashes through barriers, painfully,
> maybe even dangerously, but, uh . . . well, there it is.
> John Hammond: [*sardonically*] There it is.
> Henry Wu: You're implying that a group composed entirely of female
> animals will . . . breed?
> Dr. Ian Malcolm: No. I'm, I'm simply saying that life, uh . . . finds a way.[1]

Life finds a way. Whatever happens, life will continue, and, in this case, dinosaurs will somehow manage to continue to exist. Well, it is not as simple as that, especially as more than 99 percent of species that have ever lived on Earth are now extinct.[2] Contrary to what we might want to believe, we have no good reasons to believe that there is any inherent purpose in life; in contrast, life might even cease to exist one day. In the previous parts of this book, I showed with concrete examples that the outcomes in human development, life, and evolution are affected by the contingencies of the respective processes. Therefore, any view about plan and purpose in life and nature is closer to wishful thinking rather than a fact of life. But the design stance seems to make us prone to accept such a view. The aim of the present book has therefore been to show: (1) that notions of genetic fatalism, destiny, and intelligent design are not supported by the evidence, and (2) that outcomes in human development, human life, and human evolution have been shaped by particular turning points, some of which I have presented in detail. This implies that that there are no predetermined outcomes in any of these historical processes. However, this does not mean in any way that everything in life is random or a matter of chance; contingency is different than mere chance and randomness.

I have concluded the last chapters of parts 2, 3, and 4 of the book with a diagram illustrating the same underlying principle: among several (many, or just a few) possible outcomes, only one was materialized. This was you or me among our 70 trillion theoretically possible siblings that our parents could have had; this was *On the Origin of Species* and not any other format in which Darwin could have published his theory; this was *Homo sapiens* and not some other human species among those that have lived on Earth. In all cases, particular critical events eventually brought about these particular outcomes and not any other possible ones. These events were **turning points**, because they were contingent *per se* and all subsequent events were contingent *upon* them. Turning points are very important for all historical processes, such as human development, life, and evolution—by "historical" I mean that they consist of sequences of events that are unique in space and time. My account in all cases has been far from exhaustive; several other turning points could be identified, but I have presented only what I consider the most obvious ones.

I must address at this point a potential objection: for many of the turning points I have identified, we cannot know all of the details about why they occurred: for instance, why a particular spermatozoon of my father (and not another) made it to the ovum of my mother and the embryo from which I developed thus emerged; why Darwin really wanted to travel aboard the *Beagle*; or what enabled our ancestors to develop cultural learning and collective knowledge. However, I argue that we do not really need to know all of the details of why these turning points occurred, in order to appreciate their impact. Let me explain this via an analogy with a simple historical process. Imagine you leave open a door that has glass frames, and you leave the room. After a while, you hear the door slam, and then you hear the noise of glass breaking. When you go to see what happened, you find the door closed, with all of its glass frames broken. Therefore, you can claim that the slamming of the door was critical for the current condition of the door: its glass frames are broken. The counterfactual outcome would be that the door would have all of the frames intact, had the door not been slammed. But because the door was slammed, the glass frames are now broken. Here is the key point: the fact that the door slammed was a turning point for the breaking of the glass frames. The slamming of the door was contingent *per se*, and the breaking of the glass frames was contingent *upon* it (and, in fact, it was because of it). To claim this, you do not really need to know why the door slammed closed in the first place: was it because of the wind coming in through an open window, or because someone got angry and slammed the door? To fully explain the breaking of the glass frames, one certainly needs to figure out these causal influences. But these are not necessary in order for you to claim that the slamming of the door was a turning point for its current condition.

Similarly, in order to identify turning points in human development, life, and evolution, we do not need to know all of the related causal details. Rather, all we need to do is to appreciate the impact of these critical events, which we consider turning points, on the subsequent outcomes; that is, we need only to appreciate that the outcomes were contingent *upon* those events, as well as that these critical events were themselves contingent *per se.* This is, I argue, a common underlying principle of human development, life, and evolution: outcomes are shaped by turning points. Now, in order to better appreciate this common principle, we also need to consider the differences between these three historical processes. Overall, there are two major kinds of differences, pertaining to: (1) the kind of factors that influence the final outcomes; and (2) the epistemic certainty that it is possible to have about these factors and the sequences of events.

Let us explore the first difference. The outcomes in the cases that I have discussed in this book depend both on natural processes (and the events taking place during them) and on human decisions; however, their dependence on each of these factors varies. Human development depends a lot on how the natural processes of reproduction and embryo growth will occur in the bodies of the parents. It also depends on external conditions that may affect the physiology of these processes. Of course, human decisions also play a role; for instance, the decision of a woman to smoke during pregnancy or not to follow a healthy diet may have a negative impact on the embryo. Human evolution also depended on human decisions, for example, the decision of our ancestors to switch from arboreal life to terrestrial life or to live in communities of collective knowledge. However, neither the chromosome fusion nor the anatomic changes that led to our current form could have been the outcome of human decisions. Development and evolution are unconscious natural processes, the outcomes of which depend a lot on the initial conditions (e.g., the alleles that existed in the reproductive cells of one's parents, in the case of development; the genetic variation available within a population, in the case of evolution), and on how the natural processes take place (e.g., mutations or chromosome shuffling, in the case of development; available resources that might cause selection, in the case of evolution). In contrast, whereas natural processes and events also impact human life (e.g., Darwin could have died on the *Beagle* voyage during a storm), in the case of individual human lives one's choices and decisions can also be decisive factors. As I have shown in the case of Darwin, it was his ideas, choices, and decisions that shaped the development of his theory and for the particular outcome, *On the Origin of Species by Means of Natural Selection*, to be what it was. However, my understanding is that this difference among the three processes is one of degree rather than kind.

The second important difference among human development, life, and evolution has to do with the epistemic access and certainty that is possible to have about the respective factors and the sequences of events. As shown in the various chapters of the present book, there are different kinds of evidence on which we can rely to make conclusions about these processes. At the same time, the epistemic certainty that we can have about these historical processes is different exactly because the epistemic access that we can have to the respective events is different. Before examining this in detail for each kind of process, I must note that epistemic certainty is anyway an unattainable goal for the vast majority of inquiries. The reason for this is that we cannot really know every single detail, especially for historical processes that are unique in space and time. However, and this is what matters in my view, we do not need to have absolute certainly for understanding. Even if we do not have access to all of the details of these processes, we often have sufficient evidence to understand what happened (as illustrated by the above example of the glass frames of the door).

Epistemic access and certainty are less in human evolution than in the other two cases. The problem here is that we cannot always have sufficient and complete evidence, either because it does not exist or because we have not yet found it. For all of the major turning points described in part 4 of the present book we cannot have direct access to when, how, and why the respective changes leading to chromosome fusion, bipedalism, prolonged brain development, and biocultural evolution took place. We cannot travel back in time and observe what happened. However, all of these processes have left traces that we can examine today and draw conclusions from them about what happened. Conclusions can come only after the study of the available evidence, and interpretations may therefore vary. Thus, experts debate the details among themselves, and this is healthy for science. Uncertainty about the details is what motivates further research and eventually what makes science advance. But there is no disagreement over the broad picture of human evolution.

Things can become a little better regarding epistemic access and certainty in the case of human life, if the events are recent, like the ones of my own life that I have written about in the preface, or if there are extensive written records left, as in the case of Darwin, who left us diaries, notebooks, correspondence, and published works. It would not have been possible to describe the turning points leading to *On the Origin of Species* if Darwin had not written his early thoughts in his notebooks and diaries, if his correspondence had not been saved, or if he had not written his autobiography toward the end of his life. The latter is of special importance for the present book, as I selected the turning points on the basis of my interpretation of the significance that he attributed to them in his autobiography and of what exactly he wrote about them. When such extensive written records are available, we can be confident that we have sufficient access to the thoughts of the person, and so we can reasonably analyze this person's decisions and how the background conditions in each case influenced them. Of course, there is still room for interpretation, but in cases of such a rich written record like the one that Darwin has left, we can be confident about our conclusions.

Finally, in the case of human development, epistemic access and certainty can become even higher. Whereas we cannot repeat the developmental process of each one of us and figure out exactly what happened during it, we can study the results of failed developmental processes and conclude sufficiently well what happened or did not happen there. Indeed, there is so much that can go wrong during the various cell divisions leading to the reproductive cells, as well as those within the embryo, that we can appreciate how lucky we are to be here. In principle, we could even experiment with developmental processes, but ethically speaking this is not appropriate to do. Yet, some types of experimentation are possible, for example, when excess embryos from IVF procedures are given to researchers, when parents consent to doing this. In addition, there are several "natural" experiments also going on, as unfortunately pregnancies are naturally terminated in the case of miscarriages. When researchers study those miscarried embryos, if and whenever this is possible, important conclusions can be drawn about what went wrong in those cases, and therefore what did not go wrong in our own.

Having now clarified the differences in the kinds of factors that influence the final outcomes, and in the epistemic certainty that is possible to have about these factors and the sequences of events in human development, life, and evolution, we can now return back to the similarities. In the present book, I have focused on what I have identified as a major similarity among these processes: that critical events shape outcomes, being turning points because they are contingent *per se* and because subsequent events are contingent *upon them*. This idea becomes especially important exactly because not everything in these processes is contingent. In contrast, in human development, life, and evolution there are robust processes at work that, more or less often, lead to similar outcomes. It is exactly because of the existence of these robust processes that turning points are very significant and important. But before explaining that, let us see what these robust processes are in each case.

Organisms generally develop toward a final form that is, overall, predictable: plant seeds will develop to plants; pig embryos will develop to pigs; and human embryos will develop to humans with a head, two eyes, two ears, two arms, two legs, and so on. This happens in most cases despite perturbations inside and outside the embryo, and in general individuals consistently end up having the general characteristics of their species, more or less in the same form, no matter where they live. This phenomenon is called developmental robustness, and it happens because various processes operating at different levels, from the molecular level to that of the organism, are coordinated in such a way that the adult phenotype is produced. However, the development of organisms is also plastic enough to accommodate changes stemming from influences from the immediate environment, and thus there is variation among the members of the same species. Therefore, the adult phenotype is the outcome of the interaction of developmental robustness and developmental plasticity.[3] During this interaction, critical events can make a difference, driving development toward one of several possible outcomes. We could therefore say that *robust developmental processes can shape the outcomes of the development of humans, but the differences among them are due to turning points that eventually influence the details of their characteristics and/or disease, and whether they will be born at all.*

Regarding Darwin, one can ask the question whether the development of Darwin's theory was independent of the context of the Victorian society in which it was developed, meaning that the theory would be developed anyway, somehow and somewhere; or whether the development of that theory is in fact inseparable from the context of Victorian society and in this sense the contingent outcome of a particular social history. As I have already explained, particular characteristics of the Victorian era were crucial for the development of the theory of natural selection: natural theology; the popularity of natural history; animal and plant breeding; political economy of the kind of Malthus and Smith; the industrialization and imperialism of the British Empire, and so on. It is therefore no coincidence that Darwin developed the theory that he did, which exhibits influences of all the aforementioned features of the Victorian era; it is also no coincidence that the person who came up with a similar theory around the same time with Darwin, Alfred Russel Wallace, was British too.[4] Perhaps someone else from the same context would have developed the theory at some later time. But as I have shown, Darwin developed a distinctive theory because of particular influences during his life. We could therefore say that *robust socio-historical processes, in particular, socio-cultural contexts, can shape the development of scientific theories, but the differences among them are due to turning points that eventually influence their details and whether they will be developed at all.*

Finally, as already mentioned in the introduction, in evolution there are several similar characteristics that have independently evolved at different times in different lineages. This phenomenon is called convergence. In general, the evolution of similar characteristics can be explained on the basis of common ancestry. For example, the evolution of wings in the various species of birds is due to their similar developmental processes that result in their forelimbs having the shape of wings, in them having a relatively light skeleton, and in them being covered by feathers. Such characteristics are called homologies. However, similar characteristics can also evolve independently if they are advantageous, and they are called homoplasies. For instance, the wings of bats consist of their elongated digits, which are connected via a webbed membrane of skin and which are different from the wings of birds. In this sense, the wings of eagles and hawks are homologies, whereas the wings of eagles and bats are homoplasies. The important point here is that evolution is possible in particular directions but not in any direction; the number of evolutionary pathways available to life are not endless. At the same time, similar environmental pressures can favor particular advantageous characteristics but not others. In this sense, one might think that humans would have evolved on Earth in some way or another.[5] But as I have shown, that humans evolved toward our present condition was contingent upon particular turning points leading to a distinctive chromosome number, bipedalism, larger brains, and cultural learning. We could therefore say that *robust evolutionary processes can shape the outcome of human evolution, but particular turning points influenced which characteristics our species has and whether we would have evolved at all.*

Jonathan Losos, a biologist with a longtime interest in evolutionary contingency, put it excellently in his marvelous book *Improbable Destinies: Fate, Chance, and the Future of Evolution*, when he wrote that whereas life on other planets—were it to exist—would likely exhibit similarities with life on Earth, there would still be enormous differences between them. Similarities notwithstanding, because of the very divergent environments on diverse planets, evolution would surely follow alternative paths. Evolution does not necessarily proceed in the same manner, and the initial conditions have an influence on the outcomes. Thus, contingencies may make a vast difference, and this is more than just a hypothesis:

> All we really need to do is compare New Zealand to anywhere else in the world to see contrasting evolutionary worlds and how they've unfolded. In the greater scheme of things, birds and mammals aren't all that different. They're not only carbon-based and use DNA, but both are vertebrate animals sharing many basic aspects of biological functioning. Yet the fauna of New Zealand is strikingly different from that of Australia, the Andes, the Serengeti, and everywhere else. No one would compare New Zealand to any of these places and say that evolution has occurred in a similar way.[6]

In his book, Losos provides fascinating accounts of numerous examples of evolution in the lab and in nature. Whereas convergence is certainly occurring, this is often the case between related organisms that may therefore have a common underlying genetic background—hence the similarities. But even if convergence were the prevalent evolutionary pattern on Earth, there would still exist unique cases—like ourselves. There always exist differences between lineages evolving in a similar manner under similar conditions. These give rise to two questions: (1) What are the exact criteria for identifying convergence, and also for distinguishing it from homology? and (2) If convergence is so prevalent, why didn't any human-like species evolve anywhere else on Earth? Both of these questions are hard to answer definitively. But in many cases of both experimental evolution and evolution in nature, the initial conditions seem to have made a difference, leading to unique paths. We cannot replay the tape as Stephen Jay Gould suggested when arguing for the importance of evolutionary contingency in his book *Wonderful Life*, as we saw in the introduction; however, we can see significant differences in how the tape actually played out (so to speak) in different natural habitats and in the lab.

Perhaps the most striking example of the impact of contingency on the evolution of life of Earth is the meteorite impact that took place about 65 million years ago, which drove nonavian dinosaurs to extinction and thus opened up a space for the evolution of mammals and ourselves. As Walter Alvarez, the geologist who in the 1980s provided evidence for the meteorite impact, put it:

> It is hard to imagine that an outside observer, dropping in to visit Earth 150 or 175 million years ago, could ever had predicted that mammals, including large and intelligent ones, would ever become the dominant animals on this planet and that dinosaurs would be represented only by their descendants, the birds, most of them quite small in size. The dominance of mammals was the result of contingent rerouting of major pathways in life history.[7]

According to Alvarez, this event meets all criteria for contingency: it was rare, unpredictable, and very significant. The meteorite impact was rare because we have documented only two or three such events on Earth during the last three to four billion years, and most asteroids (and many of the less abundant comets) never come closer to the sun than as close as the Earth is, so it is highly unlikely that they would hit our planet. It was unpredictable because even if the motion of an orbiting body is determined by invariant laws, the initial conditions are less well known and so the course taken cannot be predicted. And, finally, it was significant because of the change in the course of evolution that it brought about on Earth (nonavian dinosaur extinction and mammalian dominance).[8] This view is in full agreement with the view I have presented in this book, and the meteorite impact can be viewed as nothing but a turning point in the evolution of life on Earth.

Perhaps most interestingly, the impact of contingency can be observed at the molecular level as well. It seems to be the case that the current form of particular molecules, such as the glucocorticoid receptor, a protein receptor to which cortisol and other related molecules bind, has strongly depended on improbable, nondeterministic events, due to intrinsic bio-physical properties of the protein. Both the amino-acid sequence of this protein and its form, the characteristics that underlie the function of the protein, seem to have been contingent outcomes of the mutations that occurred, which would be unlikely to occur again if the tape could be replayed. Rather, a receptor to which cortisol binds might evolve differently in the glucocorticoid receptor itself, or by an entirely different protein, or not at all.[9] This is why in order to understand why molecules turned out as they did, we need to characterize not only the path that evolution actually followed but also the many alternative paths that could have been followed. A complex analysis of the structure of a particular protein involved in the expression of genes has shown that hundreds of alternative protein sequences could have performed the same function at least as well as the one that actually occurred. As Joseph Thornton, the lead researcher in these studies, and his colleagues nicely put it in concluding their article:

> If one of the many alternative SRE-specific outcomes had instead evolved from the ancestral protein by chance, it too could have been subsequently locked in, yielding conservation and the illusion that it evolved deterministically. The singularity of the present seems to rationalize the past. History leaves no trace of the many roads it did not take, or of the possibility that evolution turned out as it did for no good reason at all.[10]

The central thesis of the present book therefore is that *robust processes guide human development, life, and evolution, but the differences in their outcomes are due to particular turning points, that is, critical events that are contingent per se and upon which all subsequent events are contingent.* This clearly follows Gould's thesis that we should think of probability and contingency as ontological properties of nature, not as consequences of our mental limitations that preclude us from understanding the deterministic nature of the world.[11] This being noted, I do not claim that contingency is more important than convergence, or that single critical events are more important than robust processes. My argument is only that when it comes to the details of human development, life, and evolution, we need to consider the impact of turning points rather than intuitively attribute the outcomes of these processes, as many people often do, to fate, destiny, or design. In other words, we should refrain from considering outcomes as predictable or inevitable, and instead try to understand the critical events that had a causal influence upon them and made them occur. In short, to acquire a better understanding of these processes and the respective outcomes, we should try to overcome the design stance. This means that we need to impose our conscious understanding upon our intuitive thinking.

But how can we overcome the design stance? Well, it's time for some philosophical speculation. To answer this question, we first need to figure out the main features of the design stance. When people think that their fate is determined by their genes, that their destiny is predetermined, and that humans are the product of an intelligent designer who determined their features, there are two common underlying ideas. One is causal determinism, which has to do with a causal factor determining an outcome (genes determining fate; an external factor determining our destiny; and God determining our biological features). The other has to do with the certainty about what exactly the future outcome will be. The stronger the determinism, the more certain the final outcome. The general scheme underlying the design stance therefore could be this: *a factor causally determines an outcome that is therefore certain*. Let's examine causal determinism and certainty a bit more.

Causal determinism is the idea that every event is the necessary consequence of earlier events and conditions under the influence of the laws of nature.[12] This means that if an event A occurs, then another event B that is a necessary consequence of A must also occur in a law-like manner. That is to say, given that A has occurred, it is impossible for B to not occur. Causal determinism suggests that everything that happens is fixed in this way. In other words, according to causal determinism, the laws of nature and the initial conditions of the universe completely fix everything that ever has happened or will happen—which might even include our thoughts and actions. It is a view of the universe under which things must happen in the way that they do—there is no possibility for things to happen differently.

Certainty has to do with our beliefs, which in turn are the basis of our knowledge. There are at least two types of certainty, psychological and epistemic. A belief is psychologically certain when the person who holds it is convinced of its truth; there is nothing that a person can believe more strongly. Ideally, but not in most cases, a person will only be psychologically certain when that person's belief is also epistemically certain. A belief is epistemically certain when it is very well justified—so well justified that there is nothing that could be learned that would make the belief any more reasonable.[13] This means that an epistemically certain belief is already so well justified that there is no point in gathering further evidence about it because there is nothing that can strengthen (or weaken) the reasonableness of the belief. Clearly, there exist very few (if any) epistemically certain beliefs.

Causal determinism and psychological and epistemic certainty are not per se bad or wrong. Actually, epistemically certain beliefs would be great, if we could have them. In fact, there also exist several simple physical processes that are causal-deterministic and the outcomes of which can be predicted and explained with certainty. Here is a simple example to illustrate this. If I hold a ball in my hand and I decide to release it, a very likely consequence is that the ball will fall. This is due to the properties of the ball (such as its weight), the condition that I no longer exercise any force to counter its weight, and that it is a law of nature that any object with weight will fall when it is released, because of the gravitational field of Earth. This outcome is psychologically certain and while it is not epistemically certain, it is about as close as we can get to that kind of certainty, because we can be convinced about its truth and because we can have a high degree of commitment to it, as it can be repeatedly tested and confirmed.

If causal determinism is true, then the more certain one is about the laws of nature and the initial conditions, the more certain one should be of the outcome. In a sense, causal determinism suggests that the laws of nature plus the initial conditions entail what will happen. This is like a deductive argument; if you are certain of the premises, you can also be certain of the conclusion. The (currently outdated) deductive-nomological account of scientific explanation actually took this form. According to that, a good scientific explanation is a deductive argument in which propositions expressing the relevant scientific law and the initial conditions logically entail that the event being explained must occur. This is an example of logical deduction.[14]

Initial conditions
Laws of nature   (what is doing the explaining)
_____

Therefore, the outcome   (what is explained)

Here is an actual example:

Ball has weight; ball is released from
    hand; there is a gravitational field
Objects with weight fall "down" with
    a gravitational field
_____

Therefore, ball falls down when I
    release it from my hand

This is accurate insofar as we think of the particular process in general terms. This is what would, in principle, happen. Indeed, if you repeated this process numerous times, most times you would end up with the same result: the ball on the ground. However, this is different from studying and understanding every one of these processes after they took place to understand why and how they occurred. Most balls would indeed reach the ground. But one ball might have fallen on the table next to you and not on the ground, because it slipped from your hand; or maybe someone next to you might have caught the ball after it was released, so it never reached the ground; or a ball might be filled with helium and fly away from you once you released it. Even though such cases might seem rare or improbable, they might occur even in this otherwise predictable, easy-to-explain physical process. However, biological processes are not characterized by a similar level of causal determinism and certainty. Let us see why.

Biological processes do not operate under invariable laws, such as those of physics. Several biological processes operate in an apparently law-like manner, for example, if there is variation and struggle for existence in a population, evolution by natural selection will most likely occur; or if an embryo is implanted on its mother's uterus, it will most likely develop and be born. But note the probabilistic dimension in these statements. It is not possible for us to be certain of the outcomes, because these outcomes are not strictly determined. In the physical process described above, the antecedent conditions were less important than the natural law. The ball will fall if it is released, and only in the exceptional cases that I mentioned above it will not reach the ground. But in biology, the particular circumstances in the antecedent conditions are those that are causally responsible and that form the basis of explanation and prediction.[15] This is what the various examples that I have described in the present book show. Human development will occur anyway, but there are particular contingencies that may affect its outcome; a theory of natural selection could have been developed anyway, but particular contingencies affected its published form and format; the evolution of human-like forms might have occurred anyway, but particular contingencies guided the evolution of our own species as it is.

Therefore, what we need to realize is that in historical processes—in the sense of being unique in space and time, such as human development, life, and evolution—there are many different causes with major or minor effects that all together affect the final outcomes. Because of this, there are many possible outcomes, and none of them is a necessary consequence of the particular antecedent conditions. Indeed, the same conditions could have more than one different outcome. Therefore, the level of certainty that we can have is limited. In a multi-causal world, we are not justified in believing that one or the other outcome will definitely occur. But after its occurrence, we may be able to explain it, if sufficient evidence is available. This explanation would have the form of a narrative that would identify past events and conditions that were critical for the observed outcome. This would also help us identify the specific contingencies of antecedent states, which would not have generated the particular outcome if they had been different. These contingencies do not diminish the explanatory power of the narrative insofar as we know enough about antecedent states in order to understand their causal relation to the observed outcome.[16]

To overcome the design stance, I suggest that we need to reject our tendency to think in causal-deterministic terms and to look for certainty in predicting and explaining outcomes. This in turn requires us to realize that we live in a multi-causal world and that we cannot always have epistemic certainty about causes and their effects. Several different causes may have the same effect, and a single cause may have several different effects. To better understand this, we need to switch to a different conception of causes than the one that we usually have, that is, of a factor that singlehandedly brings about an effect. Rather, we should think of causes as difference-makers, that is, as factors a change in which may result in a change in the observed effect. For example, genes do not alone cause our characteristics, but changes in a gene may result in a different characteristic or in a disease. A single event such as the public reaction to a book or the reception of a letter might make the difference in one's decision to refrain from publishing or to proceed to publication. A single chromosome cannot make an organism, but a change in the chromosomal constitution of an organism (like a chromosome fusion) can in the long run contribute to the evolution of a distinct species in a particular direction. For this reason, your birth, *On the Origin of Species*, and our evolution were contingent upon various critical events, some of which I have described in the preceding chapters of the present book. Those events were critical because they shaped the final outcomes, but they could not have been predicted in advance. Because they were contingent *per se*, and because the subsequent outcomes were contingent *upon* them, these critical events were **turning points**.

What are the implications of all this for our understanding of life and of ourselves? Gould's view on the topic was that:

> *Homo sapiens*, I fear, is a "thing so small" in a vast universe, a wildly improbable evolutionary event well within the realm of contingency. Make of such conclusion what you will. Some find the prospect depressing; I have always regarded it as exhilarating, and a source of both freedom and consequent moral responsibility.[17]

In contrast, Simon Conway Morris, who critiqued Gould's views on the impact of contingency, responded to Gould's views by noting that:

> We might do better to accept our intelligence as a gift, and it may be a mistake to imagine that we shall not be called to account.[18]

There are two important issues here.

The first is whether the evidence supports one or the other view. Whereas evolutionary convergence is a fact, we should not overlook those cases in which very different forms have evolved—especially on islands. Therefore, rather than generally privilege one or the other, we need to carefully examine each situation separately. For human development, it should be self-evident that contingency rules the game, given the phenomena presented in part 1 of the present book. For human life, I have given examples of the impact of contingencies from Darwin's life (part 2) and my own; with these as examples, you will most likely identify the turning points in your own life. For human evolution, there is a lot that we still do not know. But my understanding is that if we are unique, this is because of the turning points that turned human evolution to particular paths, and not because of any inherent uniqueness.

The second issue is that the conclusions that Gould and Conway Morris draw in their quotations above—as well as those that I have drawn, too—are philosophical, not scientific. There is a lot of evidence of convergence in evolution, but at the same time there is a lot of evidence that the details are the outcome of contingencies. In the present book I have used several examples to show that the details of human development, life, and evolution are the outcome of contingencies, and particular turning points. Beyond this, one can believe what one wants. Gould thought of human intelligence as the contingent outcome of evolution and suggested that we are free to make the best of it, taking the moral responsibility for our actions. Conway Morris suggested that human intelligence is a gift (given to us from someone else, apparently) that we have to appreciate because at some point we may be called into account for our actions. In my understanding, Gould concluded that we are free and that we should act responsibly; Conway Morris concluded that we should act prudently because we may be called into account.

I opt for the former.

I suggest that we do not need to live under any fear, and we do not need to hope for anything better than what we can ourselves achieve. There is no predetermined end to anticipate (besides death) or to be afraid of. The details of both the past and the future are contingent *upon* critical events that are contingent *per se*. We should therefore feel free to make the best of this life for ourselves and for others, and do good for its own sake—not because we are afraid of being called to account. This is what makes *living* different from *surviving*. We should enjoy, esteem, and value our life and existence, because it does not really seem that we were ever meant in any sense to be born, do whatever we do, or to evolve. Nobody had a detailed plan for us. Our whole existence, as well as its details, are quite fortuitous. We should therefore consider life a great opportunity, and not take it for granted, keeping in mind (appreciating or regretting) the turning points that brought us where we are. The literary writer Nikos Kazantzakis had his family write on his tomb: "I hope for nothing; I fear nothing; I am free" (Δεν ελπίζω τίποτα. Δεν φοβάμαι τίποτα. Είμαι λέφτερος).[19] I totally endorse it. Our existence is a triumph of evolution; and its details, the outcomes of contingencies. We had better appreciate it, and live it in full, for as long as it lasts.

NOTES

INTRODUCTION

1. It is far from simple to define what the "public" is, as there may be several different definitions. In the present book I use this noun vaguely to refer to all ordinary people, following the definition in Cambridge dictionary. *Cambridge Dictionary*, s.v. "Public," http://dictionary.cambridge.org/dictionary/english/public (accessed October 12, 2016).

2. Justin McCarthy, "Americans' Views on Origins of Homosexuality Remain Split," Gallup, Washington, DC, May 28, 2014, http://www.gallup.com/poll/170753/americans-views-origins-homosexuality-remain-split.aspx?gsource=genetics &gmedium=search&gcampaign=tiles; Josephine Mazzuca, "Origin of Homosexuality? Britons, Canadians Say 'Nature,'" Gallup, Washington, DC, November 2, 2004, http://www.gallup.com/poll/13930/origin-homosexuality -britons-canadians-say-nature.aspx (accessed October 20, 2016).

3. Linda Lyons, "Paranormal Beliefs Come (Super)Naturally to Some," Gallup, Washington, DC, November 1, 2005, http://www.gallup.com/poll/19558/paranormal-beliefs-come-supernaturally-some.aspx (accessed December 27, 2016).

4. "Evolution, Creationism, Intelligent Design," Gallup, Washington, DC, http://www.gallup.com/poll/21814/evolution-creationism-intelligent-design.aspx (accessed December 27, 2016).

5. For an overview of the relevant research see Kostas Kampourakis, *Understanding Evolution* (Cambridge: Cambridge University Press, 2014); Kostas Kampourakis, *Making Sense of Genes* (Cambridge: Cambridge University Press, 2017).

6. Celeste Condit, "What Is 'public opinion' about Genetics?" *Nature Reviews Genetics* 2, no. 10 (2001): 811–15; Kevin McCain and Kostas Kampourakis, "Which Question Do Polls about Evolution and Belief Really Ask, and Why Does It Matter?" *Public Understanding of Science* (2016): 0963662516642726.

7. This conception comprises three others: *genetic essentialism*, the idea that genes inside us specify who we are; *genetic determinism*, the idea that this is done notwithstanding the environment; and *genetic reductionism*, the idea that therefore if we want to understand why we are the way we are, we have to study our genes

(see Kampourakis, *Making Sense of Genes*, p. 6). In the present book, I refer to these conceptions (both individually and collectively) as *genetic fatalism.*

8. For the distinction between fate and destiny, see Richard W. Bargdill, "Fate and Destiny: Some Historical Distinctions between the Concepts," *Journal of Theoretical and Philosophical Psychology* 26, no. 1-2 (2006): 205–220; for the concept of inevitability see Simon Conway Morris, *Life's Solution: Inevitable Humans in a Lonely Universe* (Cambridge: Cambridge University Press, 2003).

9. Joseph Henrich, Steven J. Heine, and Ara Norenzayan, "Most People Are Not WEIRD," *Nature* 466, no. 7302 (2010): 29.

10. John P. A. Ioannidis, "Why Most Published Research Findings Are False," *PLoS Medicine* 2, no. 8 (2005): e124.

11. Thomas Kelly, "Evidence," in *The Stanford Encyclopedia of Philosophy*, fall 2014 ed., edited by Edward N. Zalta (Stanford, CA: Stanford University, 2014), http://plato.stanford.edu/archives/fall2014/entries/evidence/ (accessed September 12, 2017).

12. Stephen Jay Gould, *Wonderful Life: The Burgess Shale and the Nature of History* (1989; Vintage: London, 2000), p. 284.

13. Ibid., p. 48.

14. Ibid., p. 51.

15. Ibid., p. 289.

16. Simon Conway Morris, *The Crucible of Creation: The Burgess Shale and the Rise of Animals* (Oxford: Oxford University Press, 1998), p. 14.

17. For various examples of convergence in evolution, see George R. McGhee, *Convergent Evolution: Limited Forms Most Beautiful* (Cambridge, MA: MIT Press, 2011); Conway Morris, *Life's Solution.*

18. Stephen Jay Gould, *The Structure of Evolutionary Theory* (Cambridge, MA: Harvard University Press), pp. 1333–34. Gould also noted that this was a central feature in Darwin's theorizing that had been underemphasized by his followers in an attempt to win more prestige for evolution under the idea that science was explained by general laws and not by particular narration (p. 1336).

19. Ibid., p. 340. John Beatty, "Replaying Life's Tape," *Journal of Philosophy* 103, no. 7 (2006): 336–62, 340.

20. John Beatty, "What Are Narratives Good For?" *Studies in History and Philosophy of Science Part C: Studies in History and Philosophy of Biological and Biomedical Sciences* 58 (2016): 33–40.

21. For excellent and readable discussions of randomness and probabilities see Gerd Gigerenzer, *Calculated Risks: How to Know When Numbers Deceive You*

(New York: Simon & Schuster, 2002); Leonard Mlodinow, *The Drunkard's Walk: How Randomness Rules Our Lives* (London: Penguin, 2008); David J. Hand, *The Improbability Principle: Why Coincidences, Miracles, and Rare Events Happen Every Day* (New York: Farrar, Strauss, and Giroux, 2014).

22. The release of each ball is an event independent from the others. Given that there exist three possible routes, each time that a ball is released there is a probability of one out of three: 1/3 or 33.333 . . . percent) that it will take one of these routes; this is because the plane is constructed in such a way that does not bias the direction the balls will take. Therefore, when we release three balls consecutively and independently from one another, the probability of having *any* of the combinations in table I.1 is 1/3 × 1/3 × 1/3 = 1/27 (or 3.703 . . . percent). Here is where the math starts to get a little bit tricky. The probability of any two independent events to occur *consecutively* is the probability of one times the probability of the other. This entails that a combination such as ABC is not more probable than ABB or AAA. Each of these options have exactly the same probability to occur. However, there are six different ways that we can end up with the three balls each take different routes, as shown in the left column of table I.1 (6/27 or 2/9 or 22.222 . . . percent). But there are also eighteen possible combinations of two balls taking the same route (18/27 or 6/9 or 2/3 or 66.666 . . . percent). This means that having the balls take each a different route is overall less probable than having two balls taking the same route (roughly 22 percent versus 66 percent, respectively). As you can readily see, the latter is actually three times more probable than the former! Now that we have examined this analogy, the first conclusion we can make, therefore, is that probabilities can easily deceive us.

23. In this case, we are talking about conditional probabilities, which is a measure of the probability of an event to occur given that another event has previously occurred.

24. Richard J. Evans, *Altered Pasts: Counterfactuals in History* (London: Abacus, 2016), p. xi.

25. Genotype and phenotype are technical terms that are easy to understand. For the purposes of the present book, "genotype" will be used to refer to which genes or other DNA sequences a person has, whereas "phenotype" will refer to the biological features (external, such as eye color, or internal, such as blood groups) that a person exhibits. Generally speaking, genotypes often affect phenotypes, and phenotypes often depend on genotypes. However, there is never a one-to-one relation in the sense that a genotype determines a phenotype.

NOTES

CHAPTER 1: "WHY X?": "IN ORDER TO Y"

1. The Greek term is actually *telological* but *teleological* is the term that has been commonly used in English.

2. For a fascinating, accessible, and recent account of purpose, see Michael Ruse, *On Purpose* (Princeton and Oxford: Princeton University Press, 2017); for a more philosophical discussion of Aristotle, and Plato as well, see James G. Lennox, *Aristotle's Philosophy of Biology: Studies in the Origins of Life Science* (Cambridge: Cambridge University Press, 2001); for anyone interested in reading Plato and Aristotle in Greek (ancient and modern), see Βασίλης Κάλφας, *Πλάτων Τίμαιος* (Αθήνα: Εκδόσεις Πόλις, 1995) [that is, Basilis Kalfas, *Plato Timaeus* (Athens: Polis Editions, 1995)] and Βασίλης Κάλφας, *Αριστοτέλης Περί Φύσεως: Το Δεύτερο βιβλίο των Φυσικών* (Αθήνα: Εκδόσεις Πόλις, 1999) [that is, Basilis Kalfas, *Aristotle on Nature: Physics' Book II* (Athens: Polis Editions, 1999)].

3. Frank C. Keil, "The Origins of an Autonomous Biology," in *Modularity and Constraints in Language and Cognition: Minnesota Symposia on Child Psychology*, ed. M. R. Gunnar and M. Maratsos, vol. 25 (Hillsdale, NJ: Erlbaum, 1992), pp. 103–138.

4. Frank C. Keil, "The Birth and Nurturance Concepts by Domains: The Origins of Concepts of Living Things," in *Mapping the Mind: Domain Specificity in Cognition and Culture*, ed. L. A. Hirschfeld and S. Gelman (Cambridge: Cambridge University Press, 1994), pp. 234–54.

5. Deborah Kelemen, "The Scope of Teleological Thinking in Preschool Children," *Cognition* 70 (1999): 241–72. The participants were sixteen four- and five-year-old children and sixteen university undergraduates.

6. Deborah Kelemen, "Why Are Rocks Pointy? Children's Preference for Teleological Explanations of the Natural World," *Developmental Psychology* 35 (1999): 1440–52. The participants were sixteen university undergraduates, sixteen children around seven years old, sixteen children around eight years old, and sixteen children around ten years old.

7. Cara DiYanni and Deborah Kelemen, "Time to Get a New Mountain? The Role of Function in Children's Conceptions of Natural Kinds," *Cognition* 97, no. 3 (2005): 327–35. The participants were thirty-one six- and seven-year-old children and twenty-four nine- and ten-year-old children.

8. Marissa L. Greif, Deborah G. Kemler Nelson, Frank C. Keil, and Franky Gutierrez, "What Do Children Want to Know about Animals and Artifacts? Domain-Specific Requests for Information," *Psychological Science* 17, no. 6 (2006): 455–59. The participants were thirty-two four- to five-year-old children.

9. Adena Schachner, Liqi Zhu, Jing Li, and Deborah Kelemen, "Is the Bias for Function-Based Explanations Culturally Universal? Children from China Endorse Teleological Explanations of Natural Phenomena," *Journal of Experimental Child Psychology* 157 (2017): 29–48. The participants were forty-eight children across three grade levels (first, second, and fourth) and sixteen adults.

10. Deborah Kelemen and Evelyn Rosset, "The Human Function Compunction: Teleological Explanation in Adults," *Cognition* 111, no. 1 (2009): 138–43.

11. Deborah Kelemen, Joshua Rottman, and Rebecca Seston, "Professional Physical Scientists Display Tenacious Teleological Tendencies: Purpose-Based Reasoning as a Cognitive Default," *Journal of Experimental Psychology: General* 142, no. 4 (2013): 1074.

12. Daniel Kahneman, *Thinking, Fast and Slow* (New York: Farrar, Strauss, and Giroux, 2011), chap. 1.

13. Gerd Gigerenzer, *Gut Feelings: The Intelligence of the Unconscious* (London: Penguin, 2007), p. 16.

14. Ibid., pp. 42–43.

15. Daniel C. Dennett, *Intuition Pumps and Other Tools for Thinking* (London: Penguin, 2013), chap. 18.

16. Paul Bloom, *Descartes' Baby: How the Science of Child Development Explains What Makes Us Human* (New York: Basic, 2004), p. 55.

17. Richard Dawkins, *The Blind Watchmaker* (London: Penguin, 2006), p. 37.

18. Kostas Kampourakis, *Understanding Evolution* (Cambridge: Cambridge University Press, 2014); James G. Lennox, and Kostas Kampourakis, "Biological Teleology: The Need for History," in *The Philosophy of Biology: A Companion for Educators*, ed. Kostas Kampourakis (Dordrecht: Springer Netherlands, 2013), pp. 421–54.

19. If you think that all of this is complicated, things can become trickier. Imagine that one is using a sharp branch from a tree to open holes. What is this? This is indeed a tool, but it is not an artifact unless it was broken in a particular manner in order for it to be sharp. In other words, the branch is a part of an organism that was also intentionally used by someone (a human or a chimpanzee). Such natural objects have been called naturefacts, because they stand between natural objects and artifacts. Now, if the branch is further modified with an intended use in mind, it can become a genuine artifact. See R. Hilpinen, "Artifact," in *The Stanford Encyclopedia of Philosophy*, winter 2011 ed., edited by E. N. Zalta, http://plato.stanford.edu/archives/win2011/entries/artifact (accessed October 16, 2017); first published Jan. 5, 1999; substantively revised Oct. 11, 2011.

20. Frank C. Keil, "Science Starts Early," *Science* 331, no. 6020 (2011): 1022–23.

CHAPTER 2: "OUR FATE IS IN OUR GENES"

1. James D. Watson, interview by L. Jaroff, "The Gene Hunt," *Time*, March 20, 1989, pp. 62–67.

2. Walter Gilbert, "A Vision of the Grail," in *The Code of Codes: Scientific and Social Issues in the Human Genome Project*, ed. Daniel J. Kevles and Leroy Hood (Cambridge, MA: Harvard University Press, 1992), pp. 83–97, 96.

3. Lisa Gannett, "The Human Genome Project," in *The Stanford Encyclopedia of Philosophy*, summer 2016 ed., edited by Edward N. Zalta, https://plato.stanford .edu/archives/sum2016/entries/human-genome/ (accessed October 17, 2017); first published Nov. 26, 2008.

4. The official announcement of the completion of the sequencing of the human genome involved Bill Clinton, president of the United States at the time; Craig Venter, who represented Celera Genomics; and Francis Collins, who represented the public consortium to sequence the human genome. If you look at photos of the event, there is a screen behind them with the statement: "Decoding the book of life: a milestone for humanity." See, for example, Kostas Kampourakis, *Making Sense of Genes* (Cambridge: Cambridge University Press, 2017), p. 107.

5. President Bill Clinton, "Text of the White House Statements on the Human Genome Project," *New York Times*, June 27, 2000, https://partners.nytimes.com/ library/national/science/062700sci-genome-text.html (accessed December 20, 2016).

6. Polling Report, "Health Policy," p. 15, http://www.pollingreport.com/ health15.htm (accessed October 17, 2017).

7. Justin McCarthy, "Americans' Views on Origins of Homosexuality Remain Split," Gallup, Washington, DC, May 28, 2014, http://www.gallup.com/poll/170753/ americans-views-origins-homosexuality-remain-split.aspx?gsource=genetics &gmedium=search&gcampaign=tiles (accessed December 20, 2016).

8. "Divided Opinions over Why People Are Gay, Lesbian," Pew Research Center, Washington, DC, May 1–5, 2013, http://www.people-press.org/2013/06/06/ section-2-views-of-gay-men-and-lesbians-roots-of-homosexuality-personal-contact -with-gays/#divided-opinions-over-why-people-are-gay-lesbian (accessed December 20, 2016).

9. Josephine Mazzuca, "Origin of Homosexuality? Britons, Canadians Say 'Nature,'" Gallup, Washington, DC, November 2, 2004, http://www.gallup.com/ poll/13930/origin-homosexuality-britons-canadians-say-nature.aspx (accessed December 20, 2016).

10. Ravit G. Duncan and Brian J. Reiser, "Reasoning across Ontologically

Distinct Levels: Students' Understandings of Molecular Genetics," *Journal of Research in Science Teaching* 44, no. 7 (2007): 938–59. The participants in this study were students aged fifteen to sixteen.

11. Kenna R. Mills Shaw et al., "Essay Contest Reveals Misconceptions of High School Students in Genetics Content," *Genetics* 178, no. 3 (2008): 1157–68. The participants in this study were high-school students in grades 9 through 12.

12. Jenny Lewis, John Leach, and Colin Wood-Robinson, "All in the Genes? Young People's Understanding of the Nature of Genes," *Journal of Biological Education* 34, no. 2 (2000): 74–79. Participants were 383 fourteen- to sixteen-year-old students.

13. Ian Sample, "'Happy Gene' May Increase Chances of Romantic Relationships," *Guardian*, November 20, 2014, http://www.theguardian.com/science/2014/nov/20/happy-gene-romantic-relationship-serotonin-romance (accessed December 20, 2016).

14. Richard A. Friedman, "Infidelity Lurks in Your Genes," *New York Times*, May 22, 2015, http://www.nytimes.com/2015/05/24/opinion/sunday/infidelity-lurks-in -your-genes.html?partner=rss&emc=rss (accessed December 20, 2016).

15. Alexandre Morin-Chassé, "Public (Mis)understanding of News about Behavioral Genetics Research: A Survey Experiment," *BioScience* 64, no. 12 (2014): 1170–77. In this study 1,413 people were self-recruited and participated (see article for details).

16. Richard C. Lewontin, "The Analysis of Variance and the Analysis of Causes," *American Journal of Human Genetics* 26 (1974): 400–411.

17. Angelina Jolie, "My Medical Choice," *New York Times*, May 14, 2013, http://www.nytimes.com/2013/05/14/opinion/my-medical-choice.html?r=1 (accessed December 20, 2016).

18. *Daily Mail* Reporter, "Women Like Angelina Jolie Who Carry the BRCA1 Gene Are Less Likely to Die from Breast Cancer if They Have Their OVARIES Removed," *Daily Mail*, April 23, 2015, http://www.dailymail.co.uk/health/article -3052396/Women-like-Angelina-Jolie-carry-BRCA1-gene-likely-die-breast-cancer -OVARIES-removed.html (accessed September 12, 2017); Miriam Falco and Dana Ford, "Study: Women with BRCA1 Mutations Should Remove Ovaries by 35," CNN, February 25, 2014, http://edition.cnn.com/2014/02/25/health/brca-study/ (accessed September 12, 2017); Julie Revelant, "Moms with BRCA Breast Cancer Gene Mutations Face Tough Decisions," Fox News, October 18, 2015, http://www.foxnews .com/health/2015/10/18/moms-with-brca-breast-cancer-gene-mutations-face-tough -decisions.html (accessed December 20, 2016).

19. Dina L. G. Borzekowski et al., "The Angelina Effect: Immediate Reach,

Grasp, and Impact of Going Public," *Genetics in Medicine* 16 (2013): 516–21. A representative sample of 2,572 US adults participated in this study.

20. Kostas Kampourakis, *Making Sense of Genes* (Cambridge: Cambridge University Press, 2017), chap. 7 and 12.

21. Wren A. Gould, and Steven J. Heine, "Implicit Essentialism: Genetic Concepts Are Implicitly Associated with Fate Concepts," *PLoS One* 7, no. 6 (2012): e38176.

22. Ilan Dar-Nimrod, Benjamin Y. Cheung, Matthew B. Ruby, and Steven J. Heine, "Can Merely Learning about Obesity Genes Affect Eating Behavior?" *Appetite* 81 (2014): 269–76. The participants in the three reported studies were 131, 143, and 162 undergraduate students, respectively.

23. Kampourakis, *Making Sense of Genes*, p. 260.

CHAPTER 3: "EVERYTHING HAPPENS FOR A REASON"

1. Madison Park, "Small Choices, Saved Lives: Near Misses of 9/11," CNN, September 5, 2011, http://edition.cnn.com/2011/US/09/03/near.death.decisions/ (accessed December 27, 2016).

2. Linda Lyons, "Paranormal Beliefs Come (Super)Naturally to Some," Gallup, Washington, DC, November 1, 2005, http://www.gallup.com/poll/19558/ paranormal-beliefs-come-supernaturally-some.aspx (accessed December 27, 2016). The study found belief among 28 percent of women versus 23 percent of men in the United States; 33 percent of women versus 17 percent of men in Canada; and 30 percent of women versus 14 percent of men in Great Britain.

3. Lawrence M. Principe, "Myth 4. That Alchemy and Astrology Were Superstitious Pursuits That Did Not Contribute to Science and Scientific Understanding," in *Newton's Apple and Other Myths about Science*, ed. Ronald L. Numbers and Kostas Kampourakis (Cambridge MA: Harvard University Press, 2015), pp. 32–36.

4. "Horoscopes," Astrology.com, https://www.astrology.com/horoscope/daily .html (accessed December 27, 2016).

5. Donald R. Prothero, *Reality Check: How Science Deniers Threaten Our Future* (Bloomington & Indianapolis: Indiana University Press, 2013), pp. 207–21.

6. Shawn Carlson, "A Double-Blind Test of Astrology," *Nature* 318, no. 6045 (1985): 419–25.

7. Michael M. De Robertis and Paul A. Delaney, "A Survey of the Attitudes of University Students to Astrology and Astronomy," *Journal of the Royal Astronomical Society of Canada* no. 87 (1993): 34–50. Participants were first-year undergraduates,

1,122 studying arts and 383 studying science. Among arts students 42.5 percent paid some and 31.7 percent paid a lot of attention to horoscopes, compared to 46.7 percent and 18 percent of science students, respectively. In addition, 55.8 percent of arts students and 50.4 percent of science students found their horoscopes somewhat accurate, whereas 18.2 percent of arts students and 14.1 percent of science students had once or twice based conscious decisions on their horoscope. Furthermore, 43 percent of arts students and 33.7 percent of science students somewhat subscribed to the principles of astrology. Finally, 48.5 percent of arts students and 33.4 percent of science students thought of both astronomy and astrology as being science.

8. Michael M. De Robertis and Paul A. Delaney, "A Second Survey of the Attitudes of University Students to Astrology and Astronomy," *Journal of the Royal Astronomical Society of Canada* no. 94 (2000): 112–22. Participants were 1,224 arts students and 438 science students. Among arts students, 44.8 percent paid some attention to horoscopes and 31.7 percent paid a lot of attention to horoscopes, compared to 46.1 percent and 26.5 percent of science students, respectively. In addition, 61.8 percent of arts students and 56.9 percent of science students found horoscopes somewhat accurate. Furthermore, 23.4 percent of arts students and 20.3 percent of science students had based conscious decisions on horoscopes once or twice. Finally, 50.4 percent of arts students and 49.3 percent of science students somewhat subscribed to astrology, whereas 50.7 percent of arts students and 44.1 percent of science students thought of both astronomy and astrology as being science.

9. Chris Impey, Sanlyn Buxner, Jessie Antonellis, Elizabeth Johnson, and Courtney King, "A Twenty-Year Survey of Science Literacy among College Under-graduates," *Journal of College Science Teaching* no. 40 (2011): 31–37; Hannah Sugarman, Chris Impey, Sanlyn Buxner, and Jessie Antonellis, "Astrology Beliefs among Undergraduate Students," *Astronomy Education Review* no. 10 (2011): 010101.

10. Nick Allum, "What Makes Some People Think Astrology Is Scientific?" *Science Communication* 33, no. 3 (2011): 341–66.

11. "PRRI/RNS Religion News Survey, March 2011," Association of Religion Data Archives, Pennsylvania State University, University Park, PA, March 17–20, 2011, http://www.thearda.com/Archive/Files/Descriptions/PRRIRNM.asp (accessed January 13, 2017).

12. Ara Norenzayan and Albert Lee, "It Was Meant to Happen: Explaining Cultural Variations in Fate Attributions," *Journal of Personality and Social Psychology* no. 98 (2010): 702. The participants were undergraduate students; 96 were of European ancestry (51 Christians, 45 nonreligious) and 117 of mostly Chinese East Asian ancestry (45 Christians, 72 nonreligious).

13. Konika Banerjee and Paul Bloom, "Why Did This Happen to Me? Religious Believers' and Non-Believers' Teleological Reasoning about Life Events," *Cognition* 133, no. 1 (2014): 277–303. Overall, 69 percent expressed some degree of belief in fate, whereas 12 percent were neutral and 19 percent explicitly denied belief in fate. Among those believing in God, 84.8 percent expressed some degree of belief in fate, 13 percent stated that they were neutral, and only 2.2 percent denied any belief in fate. Among nonbelievers, 54.3 percent expressed some degree of belief in fate, 5.7 percent stated that they were neutral, and 40 percent denied any belief in fate. It was also found that 60.9 percent of believers agreed that they sometimes see signs in significant life events; 56.6 percent of them agreed that "everything works out for the best in the end"; and 82.5 percent agreed that there is "order in the universe." This was the case for 20 percent, 20 percent, and 45.8 percent of the nonbelievers, respectively.

14. Aiyana K. Willard and Ara Norenzayan, "Cognitive Biases Explain Religious Belief, Paranormal Belief, and Belief in Life's Purpose," *Cognition* 129, no. 2 (2013): 379–91.

15. Bethany T. Heywood and Jesse M. Bering, "'Meant to Be': How Religious Beliefs and Cultural Religiosity Affect the Implicit Bias to Think Teleologically," *Religion, Brain, & Behavior* 4, no. 3 (2014): 183–201. Participants were seventeen atheists and seventeen theists from the United States and seventeen atheists and seventeen theists from Great Britain.

16. Konika Banerjee and Paul Bloom, "'Everything Happens for a Reason': Children's Beliefs about Purpose in Life Events," *Child Development* 86, no. 2 (2015): 503–18. Participants in this study were thirty-four five- to seven-year-old children, twenty-six eight- to ten-year-old children, and thirty-four adults.

17. David W. Moore, "Three in Four Americans Believe in Paranormal," Gallup, Washington, DC, June 16, 2005, http://www.gallup.com/poll/16915/three-four -americans-believe-paranormal.aspx (accessed February 15, 2017). Participants were 1,002 US adults. Among participants, 20 percent of them believed in reincarnation; another 20 percent were not sure what to believe about it; and 59 percent stated that they did not believe in it (1 percent expressed no opinion).

18. Erlendur Haraldsson, "Popular Psychology, Belief in Life after Death and Reincarnation in the Nordic Countries, Western and Eastern Europe," *Nordic Psychology* 58, no. 2 (2006): 171–80. Data was collected in the years 1990–1993 and 1999–2002 for the European Values Survey. The mean sample size was 1,140 for the 35 European countries and 1,233 persons for the five Nordic countries. On average, approximately 22.2 percent of Western Europeans and 22.6 of people from Nordic countries believed in reincarnation. Some data from different countries: the

highest percentages were found in Switzerland (36 percent), the United Kingdom (29 percent), and Portugal (also 29 percent); and the lowest in Malta (12 percent) and Northern Ireland (17 percent). Belief in reincarnation was higher in Eastern European countries, especially Lithuania (44 percent, the highest in the whole of Europe) and Estonia (37 percent), and lowest in Eastern Germany (12 percent) and Slovenia (17 percent).

19. Ipsos, "Ipsos Global @dvisory: Supreme Being(s), the Afterlife and Evolution," April 24, 2011, https://www.ipsos.com/en-us/ipsos-global-dvisory-supreme -beings-afterlife and-evolution (accessed October 17, 2017). The study involved 18,531 adults from 23 countries. Among them, 23 percent believed in an afterlife "but not specifically in a heaven or hell"; 19 percent believed that "you go to heaven or hell"; 7 percent believed that "you are ultimately reincarnated"; and 2 percent believed in "heaven but not hell." In contrast, 23 percent stated that "you simply cease to exist," whereas another 26 percent stated that they "don't know what happens." In particular, belief in an afterlife, but not specifically in heaven or hell, was expressed by people from Mexico (40 percent), Russia (34 percent), Brazil (32 percent), India (29 percent), Canada (28 percent), and Argentina (27 percent). The belief in going on to heaven or hell after death was more common among people from Indonesia (62 percent), South Africa (52 percent), Turkey (52 percent), the United States (41 percent), and Brazil (28 percent). Most people who answered that "you simply cease to exist" were from South Korea (40 percent) and Spain (40 percent), France (39 percent), Japan (37 percent), and Belgium (35 percent). It must be noted that fewer people than what was reported above were found to believe in reincarnation, and these came from countries such as Hungary (13 percent), Brazil (12 percent), Mexico (11 percent), Japan (10 percent), Argentina (9 percent), and Australia (9 percent). Finally, most people who answered that they "don't know what happens" were from Sweden (41 percent), Germany (37 percent), Japan (37 percent), Russia (36 percent), and China (34 percent).

20. William Hasker and Charles Taliaferro, "Afterlife," in *The Stanford Encyclopedia of Philosophy*, winter 2014 ed., edited by Edward N. Zalta (Stanford, CA: Stanford University, 2014), https://plato.stanford.edu/archives/win2014/entries/ afterlife/ (accessed February 17, 2017).

21. Howard Robinson, "Dualism," in *The Stanford Encyclopedia of Philosophy*, winter 2016 ed., edited by Edward N. Zalta (Stanford, CA: Stanford University, 2016), https://plato.stanford.edu/archives/win2016/entries/dualism/ (accessed February 17, 2017).

CHAPTER 4: "GOD'S WISDOM REVEALED IN HIS CREATION"

1. So many books have been written about evolution and religion as well as about evolution and IDC that I refrain from discussing them in detail. The definitive account for the history of creationism is Ronald L. Numbers, *The Creationists: From Scientific Creationism to Intelligent Design*, expanded ed. (Cambridge MA: Harvard University Press, 2006); for criticisms of IDC, see Robert T. Pennock, *The Tower of Babel: The Evidence against the New Creationism* (Cambridge, MA: MIT Press, 2000); Massimo Pigliucci, *Denying Evolution: Creationism, Scientism, and the Nature of Science* (Sunderland, MA: Sinauer Associates, 2002); John C. Avise, *Inside the Human Genome: A Case for Non-Intelligent Design* (Oxford: Oxford University Press, 2010).

2. "Explore ICR," Institution for Creation Research, http://www.icr.org/discover (accessed January 20, 2017).

3. "Nature Reveals God's Wisdom," Institution for Creation Research, http://www.icr.org/wisdom-of-God (accessed January 20, 2017).

4. Richard Dawkins, *The Blind Watchmaker* (1986; London: Penguin, 2006), p. 5.

5. Michael J. Behe, *Darwin's Black Box: The Biochemical Challenge to Evolution* (1996; New York: Free Press, 2006).

6. Ibid., p. 39.

7. Ibid.

8. Charles Darwin, *On the Origin of Species by Means of Natural Selection* (London: John Murray, 1859), p. 189.

9. Behe, *Darwin's Black Box*, p. 42.

10. Ibid., pp. 42–43.

11. Ibid., p. 73.

12. William A. Dembski, "Irreducible Complexity Revisited," revised February 23, 2004, https://billdembski.com/documents/2004.01.Irred_Compl_Revisited.pdf (accessed October 2, 2017), p. 43.

13. William Paley, *Natural Theology or Evidence of the Existence and Attributes of the Deity, Collected from the Appearances of Nature* (1802; Oxford: Oxford University Press, 2006), pp. 7–8.

14. Adam R. Shapiro, "William Paley's Lost 'Intelligent Design'" *History and Philosophy of the Life Sciences* 31, no. 1 (2009): 55–77.

15. Art Swift, "In US, Belief in Creationist View of Humans at New Low," Gallup, Washington, DC, May 22, 2017, http://www.gallup.com/poll/210956/belief-creationist-view-humans-new-low.aspx (accessed June 30, 2017).

16. Frank Newport, "In US, 42% Believe Creationist View of Human Origins," Gallup, Washington, DC, June 2, 2014, http://www.gallup.com/poll/170822/believe -creationist-view-human-origins.aspx (accessed June 30, 2017).

17. "Small Businesses Top Favorability List among Americans," Ipsos, August 23, 2017, https://www.ipsos.com/en-us/news-polls/favorability-poll-2017-08-23 (accessed May 30, 2013); Kostas Kampourakis and Bruno J. Strasser, "The Evolutionist, the Creationist, and the 'Unsure': Picking up the Wrong Fight?" *International Journal of Science Education Part B: Communication and Public Engagement* 5, no. 3 (2015): 271–75.

18. Cosima Rughiniş, "A Lucky Answer to a Fair Question: Conceptual, Methodological, and Moral Implications of Including Items on Human Evolution in Scientific Literacy Surveys," *Science Communication* 33, no. 4 (2011): 501–32.

19. Fern Elsdon-Baker, "Creating Creationists: The Influence of 'Issues Framing' on Our Understanding of Public Perceptions of Clash Narratives between Evolutionary Science and Belief," *Public Understanding of Science* 24, no. 4 (2015): 422–39.

20. Cary Funk and Becka A. Alper, "Strong Role of Religion in Views about Evolution and Perceptions of Scientific Consensus," Pew Research Center, Washington, DC, October 22, 2015, http://www.pewinternet.org/2015/10/22/strong-role-of -religion-in-views-about-evolution-and-perceptions-of-scientific-consensus/ (accessed January 20, 2017); Newport, "In US, 42% Believe Creationist View"; Kevin McCain and Kostas Kampourakis, "Which Question Do Polls about Evolution and Belief Really Ask, and Why Does It Matter?" *Public Understanding of Science* (2016).

21. Margaret E. Evans, "Cognitive and Contextual Factors in the Emergence of Diverse Belief Systems: Creation versus Evolution," *Cognitive Psychology* 42, no. 3 (2001): 217–66.

22. Deborah Kelemen, "British and American Children's Preferences for Teleo-Functional Explanations of the Natural World," *Cognition* 88, no. 2 (2003): 201–21; Deborah Kelemen, "Are Children 'Intuitive Theists'? Reasoning about Purpose and Design in Nature," *Psychological Science* 15, no. 5 (2004): 295–301.

23. Deborah Kelemen and C. DiYanni, "Intuitions about Origins: Purpose and Intelligent Design in Children's Reasoning about Nature," *Journal of Cognition and Development* 6, no. 1 (2005): 3–31. The participants in this study were thirty-one six- to seven-year-olds and twenty-four nine- to ten-year-olds.

24. Elisa Järnefelt, Caitlin F. Canfield, and Deborah Kelemen, "The Divided Mind of a Disbeliever: Intuitive Beliefs about Nature as Purposefully Created among Different Groups of Non-Religious Adults," *Cognition* 140 (2015): 72–88.

25. John H. McDonald, "A Reducibly Complex Mousetrap," John H. McDonald, University of Delaware, last revised January 13, 2003, http://udel.edu/~mcdonald/oldmousetrap.html (accessed February 3, 2017).

26. John H. McDonald, "A Reducibly Complex Mousetrap," John H. McDonald, University of Delaware, last updated March 14, 2011, http://udel.edu/~mcdonald/mousetrap.html (accessed February 3, 2017).

27. Kenneth R. Miller, *Only a Theory: Evolution and the Battle for America's Soul* (New York: Penguin, 2008), pp. 53–62.

CHAPTER 5: YOUR 70 TRILLION POSSIBLE SIBLINGS

1. The cells are given different names during the various stages of the meiotic division. Before the division we have oogonia and spermatogonia. During the division we have primary and secondary oocytes and spermatocytes. To avoid confusion, I am referring to these cells as ova and spermatozoa (singular: ovum and spermatozoon) independently of their stage.

2. World Health Organization, *WHO Laboratory Manual for the Examination and Processing of Human Semen*, 5th ed. (Geneva: WHO Press, 2010).

3. Simon Meyers, Leonardo Bottolo, Colin Freeman, Gil McVean, and Peter Donnelly, "A Fine-Scale Map of Recombination Rates and Hotspots across the Human Genome," *Science* 310, no. 5746 (2005): 321–24; Graham Coop and Molly Przeworski, "An Evolutionary View of Human Recombination," *Nature Reviews Genetics* 8, no. 1 (2007): 23–34.

4. Complexity increases further by changes that happen within these chromosomes before they come together in the fertilized ovum, but we can ignore this for now (more on this in the next two chapters).

5. The only exception are monozygotic twins, that is, twins emerging from the same fertilized ovum, even though differences can eventually exist between them as well—see chapter 6.

6. For how monozygotic twins come to be, see Jamie A. Davies, *Life Unfolding: How the Human Body Creates Itself* (Oxford: Oxford University Press, 2014), pp. 27, 30, 43.

7. For accessible accounts of epigenetics, see David S. Moore, *The Developing Genome: An Introduction to Behavioral Epigenetics* (Oxford: Oxford University Press, 2015); Nessa Carey, *The Epigenetics Revolution: How Modern Biology Is Rewriting Our Understanding of Genetics, Disease, and Inheritance* (New York: Columbia University Press, 2012).

8. Edith Heard and Robert A. Martienssen, "Transgenerational Epigenetic Inheritance: Myths and Mechanisms," *Cell* 157, no. 1 (2014): 95–109.

CHAPTER 6: THE APPLE MIGHT FALL
FAR FROM THE TREE

1. Kostas Kampourakis, *Making Sense of Genes* (Cambridge: Cambridge University Press, 2017).

2. Richard A. Sturm and Tony N. Frudakis, "Eye Colour: Portals into Pigmentation Genes and Ancestry," *TRENDS in Genetics* 20, no. 8 (2004): 327–32; Richard A. Sturm and Mats Larsson, "Genetics of Human Iris Colour and Patterns," *Pigment Cell & Melanoma Research* 22, no. 5 (2009): 544–62.

3. Sturm and Larsson, "Genetics of Human Iris Colour."

4. Désirée White and Montserrat Rabago-Smith, "Genotype–Phenotype Associations and Human Eye Color," *Journal of Human Genetics* 56, no. 1 (2011): 5–7.

5. Using different methods, studies have estimated the mutation rate in humans to be 1.2×10^{-8} per nucleotide per generation and $1.4 - 2.3 \times 10^{-8}$ mutations per nucleotide per generation. Given that we have 6 billion (6×10^9) nucleotides in our cells, if you do the math you can estimate a number of novel mutations (e.g., $1.2 \times 10^{-8} \times 6 \times 10^9 = 72$; $2.3 \times 10^{-8} \times 6 \times 10^9 = 138$); see Augustine Kong, Michael L. Frigge, Gisli Masson, Soren Besenbacher, Patrick Sulem, Gisli Magnusson, Sigurjon A. Gudjonsson, et al., "Rate of de novo Mutations and the Importance of Father's Age to Disease Risk," *Nature* 488, no. 7412 (2012): 471–75; James X. Sun, Agnar Helgason, Gisli Masson, Sigríður Sunna Ebenesersdóttir, Heng Li, Swapan Mallick, Sante Gnerre, et al., "A Direct Characterization of Human Mutation Based on Microsatellites," *Nature Genetics* 44, no. 10 (2012): 1161–65.

6. For more details account see Kampourakis, *Making Sense of Genes*, chap. 6. Other parameters are also important. One is the kind of cells in which a mutation occurs; for example, having a single nonfunctional blood cell might be different than having a single nonfunctional brain cell. In addition, the timing of the change is also important, as the earlier a mutation occurs during development, the more cells will inherit it and thus the more likely it is to have an effect.

7. David Turbert, "Heterochromia," American Academy of Ophthalmology, February 3, 2017, https://www.aao.org/eye-health/diseases/what-is-heterochromia (accessed March 5, 2017).

8. Edward S. Weiss and Murray L. Janower, "The Frequency of Iris Bicolor (Segmentary Heterochromia) in a School Population," *Annals of Human Genetics* 29, no. 3 (1966): 305–308; O. Stelzer, "Iris Heterochromia: Variations in Form, Age Changes, Sex Dimorphism," *Anthropologischer Anzeiger; Bericht über die Biologisch-Anthropologische Literatur* 37, no. 2 (1979): 107–116.

9. "Eye Don't Believe It! These Celebs Have Different-Colored Eyes," Fox News, http://www.foxnews.com/entertainment/slideshow/2013/05/29/eye-dont-believe-it -these-celebs-have-different-colored-eyes.html#/slide/1 (accessed March 5, 2017).

10. Abdel Abdellaoui, Erik A. Ehli, Jouke-Jan Hottenga, Zachary Weber, Hamdi Mbarek, Gonneke Willemsen, Toos Van Beijsterveldt, et al., "CNV Concordance in 1,097 MZ Twin Pairs," *Twin Research and Human Genetics* 18, no. 1 (2015): 1–12.

11. Mario F. Fraga, Esteban Ballestar, Maria F. Paz, Santiago Ropero, Fernando Setien, Maria L. Ballestar, Damia Heine-Suñer, et al., "Epigenetic Differences Arise during the Lifetime of Monozygotic Twins," *Proceedings of the National Academy of Sciences of the United States of America* 102, no. 30 (2005): 10604–609.

12. Xinxian Deng, Joel B. Berletch, Di K. Nguyen, and Christine M. Disteche, "X Chromosome Regulation: Diverse Patterns in Development, Tissues and Disease," *Nature Reviews Genetics* 15, no. 6 (2014): 367–78.

13. Hengmi Cui, Patrick Onyango, Sheri Brandenburg, Yiqian Wu, Chih-Lin Hsieh, and Andrew P. Feinberg, "Loss of Imprinting in Colorectal Cancer Linked to Hypomethylation of H19 and IGF2," *Cancer Research* 62, no. 22 (2002): 6442–46; Jeannie T. Lee and Marisa S. Bartolomei, "X-Inactivation, Imprinting, and Long Noncoding RNAs in Health and Disease," *Cell* 152, no. 6 (2013): 1308–323; Gudrun E. Moore, Miho Ishida, Charalambos Demetriou, Lara Al-Olabi, Lydia J. Leon, Anna C. Thomas, Sayeda Abu-Amero, et al., "The Role and Interaction of Imprinted Genes in Human Fetal Growth," *Philosophical Transactions of the Royal Society B* 370, no. 1663 (2015): 20140074.

14. Susanne R. de Rooij, Hans Wouters, Julie E. Yonker, Rebecca C. Painter, and Tessa J. Roseboom, "Prenatal Undernutrition and Cognitive Function in Late Adulthood," *Proceedings of the National Academy of Sciences* 107, no. 39 (2010): 16881–86.

15. Tessa Roseboom, Susanne de Rooij, and Rebecca Painter, "The Dutch Famine and Its Long-Term Consequences for Adult Health," *Early Human Development* 82, no. 8 (2006): 485–91.

16. Bastiaan T. Heijmans, Elmar W. Tobi, Aryeh D. Stein, Hein Putter, Gerard J. Blauw, Ezra S. Susser, P. Eline Slagboom, and L. H. Lumey, "Persistent Epigenetic Differences Associated with Prenatal Exposure to Famine in Humans," *Proceedings of*

the National Academy of Sciences 105, no. 44 (2008): 17046–49; see also Elmar W. Tobi, Jelle J. Goeman, Ramin Monajemi, Hongcang Gu, Hein Putter, Yanju Zhang, Roderick C. Slieker, et al., "DNA Methylation Signatures Link Prenatal Famine Exposure to Growth and Metabolism," *Nature Communications* 5 (2014): 5592.

17. David S. Moore, *The Developing Genome: An Introduction to Behavioral Epigenetics* (Oxford: Oxford University Press, 2015).

CHAPTER 7: IT COULD BE HEREDITY, IT COULD BE BAD LUCK

1. John B. S. Haldane, "Blood Royal," *Living Age* 356 (1939): 26–31.

2. Ibid.

3. Richard Stevens, "The History of Haemophilia in the Royal Families of Europe," *British Journal of Haematology* 105, no. 1 (1999): 25–32; Evgeny I. Rogaev, Anastasia P. Grigorenko, Gulnaz Faskhutdinova, Ellen L. W. Kittler, and Yuri K. Moliaka, "Genotype Analysis Identifies the Cause of the 'Royal Disease,'" *Science* 326, no. 5954 (2009): 817; Nathalie Lannoy and Cedric Hermans, "The 'Royal Disease'— Haemophilia A or B? A Haematological Mystery Is Finally Solved," *Haemophilia* 16, no. 6 (2010): 843–47.

4. Kostas Kampourakis, *Making Sense of Genes* (Cambridge: Cambridge University Press, 2017), pp. 131–35.

5. Daniel G. MacArthur, Suganthi Balasubramanian, Adam Frankish, Ni Huang, James Morris, Klaudia Walter, Luke Jostins, et al., "A Systematic Survey of Loss-of-Function Variants in Human Protein-Coding Genes," *Science* 335, no. 6070 (2012): 823–28; Patrick Sulem, Hannes Helgason, Asmundur Oddson, Hreinn Stefansson, Sigurjon A. Gudjonsson, Florian Zink, Eirikur Hjartarson, et al., "Identification of a Large Set of Rare Complete Human Knockouts," *Nature Genetics* 47, no. 5 (2015): 448–52.

6. Rong Chen, Lisong Shi, Jörg Hakenberg, Brian Naughton, Pamela Sklar, Jianguo Zhang, Hanlin Zhou, et al., "Analysis of 589,306 Genomes Identifies Individuals Resilient to Severe Mendelian Childhood Diseases," *Nature Biotechnology* (2016): 34, 531–38.

7. Douglas Hanahan and Robert A. Weinberg, "The Hallmarks of Cancer," *Cell* 100, no. 1 (2000): 57–70; Douglas Hanahan and Robert A. Weinberg, "Hallmarks of Cancer: The Next Generation," *Cell* 144, no. 5 (2011): 646–74.

8. Song Wu, Scott Powers, Wei Zhu, and Yusuf A. Hannun, "Substantial Contribution of Extrinsic Risk Factors to Cancer Development," *Nature* 529, no. 7584 (2016): 43–47; Bert Vogelstein, Nickolas Papadopoulos, Victor E. Velculescu, Shibin Zhou, Luis A. Diaz, and Kenneth W. Kinzler, "Cancer Genome Landscapes," *Science* 339, no. 6127 (2013): 1546–58.

9. Cristian Tomasetti and Bert Vogelstein, "Variation in Cancer Risk among Tissues Can Be Explained by the Number of Stem Cell Divisions," *Science* 347, no. 6217 (2015): 78–81.

CHAPTER 8: THOSE WHO DID NOT MAKE IT

1. For an overview, see Raj Rai and Lesley Regan, "Recurrent Miscarriage," *Lancet* 368, no. 9535 (2006): 601–611.

2. Ruth C. Fretts, Julie Schmittdiel, Frances H. McLean, Robert H. Usher, and Marlene B. Goldman, "Increased Maternal Age and the Risk of Fetal Death," *New England Journal of Medicine* 333, no. 15 (1995): 953–57.

3. Anne-Marie Nybo Andersen, Jan Wohlfahrt, Peter Christens, Jørn Olsen, and Mads Melbye, "Maternal Age and Fetal Loss: Population-Based Register Linkage Study," *British Medical Journal* 320, no. 7251 (2000): 1708–712.

4. Merel M. J. van den Berg, Merel C. van Maarle, Madelon van Wely, and Mariëtte Goddijn, "Genetics of Early Miscarriage," *Biochimica et Biophysica Acta (BBA)-Molecular Basis of Disease* 1822, no. 12 (2012): 1951–59.

5. Terry Hassold, Heather Hall, and Patricia Hunt, "The Origin of Human Aneuploidy: Where We Have Been, Where We Are Going," *Human Molecular Genetics* 16, no. R2 (2007): R203–208.

6. For details of these mechanisms and of how mistakes in the meiotic divisions may occur, see: Terry Hassold and Patricia Hunt, "To Err (Meiotically) Is Human: The Genesis of Human Aneuploidy," *Nature Reviews Genetics* 2, no. 4 (2001): 280–91; So I. Nagaoka, Terry J. Hassold, and Patricia A. Hunt, "Human Aneuploidy: Mechanisms and New Insights into an Age-Old Problem," *Nature Reviews Genetics* 13, no. 7 (2012): 493–504; Mary Herbert, Dimitrios Kalleas, Daniel Cooney, Mahdi Lamb, and Lisa Lister, "Meiosis and Maternal Aging: Insights from Aneuploid Oocytes and Trisomy Births," *Cold Spring Harbor Perspectives in Biology* 7, no. 4 (2015): a017970.

7. Elpida Fragouli, Samer Alfarawati, Katharina Spath, Souraya Jaroudi, Jonas Sarasa, Maria Enciso, and Dagan Wells, "The Origin and Impact of Embryonic Aneuploidy," *Human Genetics* 132, no. 9 (2013): 1001–1013.

8. Fretts et al., "Increased Maternal Age."

9. Birgit Gellersen, Ivo A. Brosens, and Jan J. Brosens, "Decidualization of the Human Endometrium: Mechanisms, Functions, and Clinical Perspectives," *Seminars in Reproductive Medicine* 25, no. 6 (2007): 445–53.

10. Gijs Teklenburg, Madhuri Salker, Mariam Molokhia, Stuart Lavery, Geoffrey Trew, Tepchongchit Aojanepong, Helen J. Mardon, et al., "Natural Selection of Human Embryos: Decidualizing Endometrial Stromal Cells Serve as Sensors of Embryo Quality upon Implantation," *PLoS One* 5, no. 4 (2010): e10258; see also G. Teklenburg, Madhuri Salker, Cobi Heijnen, Nick S. Macklon, and Jan J. Brosens, "The Molecular Basis of Recurrent Pregnancy Loss: Impaired Natural Embryo Selection," *Molecular Human Reproduction* 16, no. 12 (2010): 886–95.

11. Evelyne Vanneste, Thierry Voet, Cédric Le Caignec, Michele Ampe, Peter Konings, Cindy Melotte, Sophie Debrock, et al., "Chromosome Instability Is Common in Human Cleavage-Stage Embryos," *Nature Medicine* 15, no. 5 (2009): 577–83.

12. Jannie van Echten-Arends, Sebastiaan Mastenbroek, Birgit Sikkema-Raddatz, Johanna C. Korevaar, Maas Jan Heineman, Fulco van der Veen, and Sjoerd Repping, "Chromosomal Mosaicism in Human Preimplantation Embryos: A Systematic Review," *Human Reproduction Update* 17, no. 5 (2011): 620–27.

CHAPTER 9: THE MOST IMPORTANT EVENT IN LIFE

1. This example comes from a book that is a magnificently readable introduction to epistemology: Jennifer Nagel, *Knowledge: A Very Short Introduction* (Oxford: Oxford University Press, 2014), pp. 2–4.

2. Charles R. Darwin, *On the Origin of Species by Means of Natural Selection* (London: John Murray, 1859). The full title of the book was: *On the Origin of Species by Means of Natural Selection, or the Preservation of Favoured Races in the Struggle for Life*. In this and the next chapters I refer to this book simply as *On the Origin of Species*.

3. Peter J. Bowler, *Darwin Deleted: Imagining a World without Darwin* (Chicago: University of Chicago Press, 2013).

4. Alfred Russel Wallace, *Contributions to the Theory of Natural Selection: A Series of Essays* (New York: Macmillan, 1870), pp. iv–v.

5. Richard J. Evans, *Altered Pasts: Counterfactuals in History* (London: Abacus, 2016), chap. 2.

6. Nora Barlow, *The Autobiography of Charles Darwin 1809–1882* (1959; New York: W. W. Norton, 1969), p. 43.

7. Ibid., p. 53.

8. Ibid., pp. 54–55.

9. Ibid., pp. 57–58.

10. Ibid., p. 59.

11. Ibid., pp. 60–61.

12. Ibid., p. 84.

13. Jonathan Hodge, "Darwin, the Galápagos, and His Changing Thoughts about Species Origins: 1835–1837," *Proceedings of the California Academy of Sciences* 61, no. 2 (2010): 89–106.

14. Barlow, *Autobiography of Charles Darwin*, p. 59.

15. Ibid., p. 73.

16. Ibid., p. 61.

17. Ibid., p. 64.

18. Ibid., pp. 65–66.

19. Ibid., p. 67.

20. Thomas F. Glick and David Kohn, ed., *Charles Darwin, On Evolution: The Development of the Theory of Natural Selection* (Indianapolis/Cambridge: Hackett, 1996), p. xv.

21. Darwin, *Origin of Species*, p. 1.

CHAPTER 10: A THEORY BY WHICH TO WORK

1. Erasmus Darwin, *Zoonomia; or, The Laws of Organic Life* (London: J. Johnson, 1794), https://archive.org/details/zoonomiaorlawsof1794darw (accessed September 13, 2017); for the life and work of Erasmus Darwin see Desmond King-Hele, *Erasmus Darwin: A Life of Unequalled Achievement* (London: Giles de la Mare, 1999).

2. Nora Barlow, *The Autobiography of Charles Darwin 1809–1882* (New York: W. W. Norton, 1969), p. 43.

3. C. R. Darwin, Notebook B: [Transmutation of species (1837–1838)]. CUL-DAR121.- Transcribed by Kees Rookmaaker. (Darwin Online, http://darwin-online.org.uk/). (Direct link: http://darwin-online.org.uk/content/frameset?pageseq=1&itemID=CUL-DAR121.-&viewtype=text.)

4. Ibid., p. 21.

5. Ibid., pp. 38–39.

6. Ibid., pp. 46–47.

7. Ibid., pp. 101–102.

8. Ibid., pp. 227–29.

9. C. R. Darwin, Notebook C: [Transmutation of species (1838.02-1838.07)]. CUL-DAR122.- Transcribed by Kees Rookmaaker. (Darwin Online, http://darwin-online.org.uk/). (Direct link: http://darwin-online.org.uk/content/frameset?itemID=CUL-DAR122.-&viewtype=text&pageseq=1.)

10. Ibid., pp. 133–34.

11. For more details on Darwin and artificial selection see Michael Ruse, "Charles Darwin and Artificial Selection," *Journal of the History of Ideas* 36, no. 2 (1975): 339–50; John F. Cornell, "Analogy and Technology in Darwin's Vision of Nature," *Journal of the History of Biology* 17, no. 3 (1984): 303–44; L. T. Evans, "Darwin's Use of the Analogy between Artificial and Natural Selection," *Journal of the History of Biology* 17, no. 1 (1984): 113–40; James A. Secord, "Darwin and the Breeders: A Social History," in *The Darwinian Heritage*, ed. D. Kohn (Princeton: Princeton University Press, 1985): 519–42.

12. Darwin, C. R. Notebook D: [Transmutation of species (7-10.1838)]. CUL-DAR123.- Transcribed by Kees Rookmaaker. (*Darwin Online*, http://darwin-online .org.uk/), pp. 134–35. (Direct link: http://darwin-online.org.uk/content/frameset ?viewtype=text&itemID=CUL-DAR123.-&pageseq=1.)

13. Barlow, *Autobiography of Charles Darwin*, pp. 98–99; Darwin seems to have finished reading Malthus on October 3, 1838; the date of the reference to Malthus in his notebook D is September 28.

14. For more details on Darwin and Malthus see Peter Vorzimmer, "Darwin, Malthus, and the Theory of Natural Selection," *Journal of the History of Ideas* 30, no. 4 (1969): 527–42; Peter J. Bowler, "Malthus, Darwin, and the Concept of Struggle," *Journal of the History of Ideas* (1976): 631–50; James G. Lennox and Bradley E. Wilson, "Natural Selection and the Struggle for Existence," *Studies in History and Philosophy of Science Part A* 25, no. 1 (1994): 65–80; David Kohn, "Darwin's Keystone: The Principle of Divergence," in *The Cambridge Companion to the "Origin of Species,"* ed. M. Ruse and R. J. Richards (Cambridge: Cambridge University Press, 2009), pp. 87–108.

15. Michael Ruse, "Darwin's Debt to Philosophy: An Examination of the Influence of the Philosophical Ideas of John F. W. Herschel and William Whewell on the Development of Charles Darwin's Theory of Evolution," *Studies in the History and Philosophy of Science* no. 6 (1975): 159–81; Lennox and Wilson, "Natural Selection and the Struggle for Existence," pp. 65–80; David L. Hull "Darwin's Science and Victorian Philosophy of Science," in *The Cambridge Companion to Darwin*, 2nd ed.,

ed. J. Hodge and G. Radick (Cambridge: Cambridge University Press, 2009), pp. 173–96.

    16. Dov Ospovat, *The Development of Darwin's Theory: Natural History, Natural Theology, and Natural Selection, 1838–1859* (Cambridge: Cambridge University Press, 1981), pp. 33–38; Kostas Kampourakis, *Understanding Evolution* (Cambridge: Cambridge University Press, 2014), pp. 111–12.

    17. Janet Browne, *Charles Darwin: Voyaging* (London: Pimlico, 2003), pp. 384–86.

    18. Charles R. Darwin, *On the Origin of Species by Means of Natural Selection* (London: John Murray, 1859), pp. 4–5.

CHAPTER 11: LIKE CONFESSING A MURDER

    1. Darwin, C. R. *Notebook E*: [Transmutation of species (10.1838-7.1839)] CUL-DAR124.- Transcribed by Kees Rookmaaker. (*Darwin Online*, http://darwin-online.org.uk/). (Direct link: http://darwin-online.org.uk/content/frameset ?itemID=CUL-DAR124.-&viewtype=text&pageseq=1.)

    2. Ibid., pp. 71–72.

    3. Ibid., p. 118.

    4. Ibid., pp. 136–37.

    5. Nora Barlow, *The Autobiography of Charles Darwin 1809–1882* (1958; New York: W. W. Norton, 1969), p. 99.

    6. Charles Lyell, *Principles of Geology*, vol. 2 (London: John Murray, 1832), p. 8.

    7. Darwin Correspondence Project, "Letter no. 729," accessed on 1 May 2017, http://www.darwinproject.ac.uk/DCP-LETT-729.

    8. Darwin Correspondence Project, "Letter no. 734," accessed on 1 May 2017, http://www.darwinproject.ac.uk/DCP-LETT-734.

    9. Darwin Correspondence Project, "Letter no. 761," accessed on 1 May 2017, http://www.darwinproject.ac.uk/DCP-LETT-761.

    10. Robert Chambers, *Vestiges of the Natural History of Creation* (London: John Churchill, 1844).

    11. James A. Secord, *Victorian Sensation: The Extraordinary Publication, Reception, and Secret Authorship of Vestiges of the Natural History of Creation* (Chicago: University of Chicago Press, 2003), pp. 1–3.

    12. Darwin Correspondence Project, "Letter no. 789," accessed on 10 May 2017, http://www.darwinproject.ac.uk/DCP-LETT-789.

13. Darwin Correspondence Project, "Letter no. 804," accessed on 10 May 2017, http://www.darwinproject.ac.uk/DCP-LETT-804.

14. Darwin Correspondence Project, "Letter no. 814," accessed on 10 May 2017, http://www.darwinproject.ac.uk/DCP-LETT-814.

15. Secord, *Victorian Sensation*, pp. 431–33.

16. Barlow, *Autobiography of Charles Darwin*, p. 99.

17. Ibid., p. 96.

18. David Kohn, "Darwin's Keystone: The Principle of Divergence," in *The Cambridge Companion to the "Origin of Species*," ed. M. Ruse and R. J. Richards (Cambridge: Cambridge University Press, 2009), pp. 87–108; Alan C. Love, "Darwin and Cirripedia Prior to 1846: Exploring the Origins of the Barnacle Research," *Journal of the History of Biology* no. 35 (2002): 251–89; Dov Ospovat, *The Development of Darwin's Theory: Natural History, Natural Theology, and Natural Selection, 1838–1859* (Cambridge: Cambridge University Press, 1981).

19. See Ospovat, *Development of Darwin's Theory*; Robert J. Richards, *The Meaning of Evolution: The Morphological Construction and Ideological Reconstruction of Darwin's Theory* (Chicago: University of Chicago Press, 1992).

20. Charles R. Darwin, *On the Origin of Species by Means of Natural Selection* (London: John Murray, 1859), p. 112.

CHAPTER 12: THE MOST STRIKING COINCIDENCE EVER

1. Robert C. Stauffer, ed., *Charles Darwin's Natural Selection: Being the Second Part of His Big Species Book Written From 1856 to 1858* (Cambridge: Cambridge University Press, 1975).

2. Nora Barlow, *The Autobiography of Charles Darwin 1809–1882* (1958; New York: W. W. Norton, 1969), pp. 99–100.

3. Darwin Correspondence Project, "Letter no. 2285," accessed on 11 May 2017, http://www.darwinproject.ac.uk/DCP-LETT-2285.

4. Barlow, *Autobiography of Charles Darwin 1809–1882*, p. 100.

5. John Van Wyhe and Kees Rookmaaker, "A New Theory to Explain the Receipt of Wallace's Ternate Essay by Darwin in 1858," *Biological Journal of the Linnean Society* 105, no. 1 (2012): 249–52.

6. Barlow, *Autobiography of Charles Darwin 1809–1882*, p. 100.

7. Charles R. Darwin, *On the Origin of Species by Means of Natural Selection* (London: John Murray, 1859), p. 6.

8. Darwin, *Origin of Species*, p. 112.

9. Ibid., pp. 352–53.

10. Gregory Radick, "Is the Theory of Natural Selection Independent of Its History?" in *The Cambridge Companion to Darwin*, ed. J. Hodge and G. Radick (Cambridge: Cambridge University Press, 2003), pp. 143–67.

11. Michael Ruse, "Myth 12. That Wallace's and Darwin's Explanations of Evolution Were Virtually the Same," in *Newton's Apple and Other Myths about Science*, ed. Ronald N. Numbers and Kostas Kampourakis (Cambridge, MA: Harvard University Press, 2015), pp. 96–102.

12. Barlow, *Autobiography of Charles Darwin*, p. 102.

CHAPTER 13: THOSE TWO FUSED CHROMOSOMES

1. Nora Barlow, *The Autobiography of Charles Darwin 1809–1882* (1958; New York: W. W. Norton, 1969), p. 77.

2. Charles R. Darwin, *The Descent of Man, and Selection in Relation to Sex* (London: John Murray, 1871), pp. 2–3.

3. Eugene E. Harris, *Ancestors in Our Genome: The New Science of Human Evolution* (Oxford: Oxford University Press, 2015), chap. 2–3.

4. Jonathan Marks, *Tales of the Ex-Apes: How We Think about Human Evolution* (Oakland, CA: University of California Press, 2015), pp. 107–113.

5. Carol E. Cleland, "Methodological and Epistemic Differences between Historical Science and Experimental Science," *Philosophy of Science* 69, no. 3 (2002): 474–96; Carol E. Cleland, "Prediction and Explanation in Historical Natural Science," *British Journal for the Philosophy of Science* 62, no. 3 (2011): 551–82; Patrick Forber, and Eric Griffith, "Historical Reconstruction: Gaining Epistemic Access to the Deep Past," *Philosophy and Theory in Biology* 3, e203 (2011), DOI:10:3998/ptb.6959004.0003.003.

6. Ajit Varki and Tasha K. Altheide, "Comparing the Human and Chimpanzee Genomes: Searching for Needles in a Haystack," *Genome Research* 15, no. 12 (2005): 1746–58.

7. The Chimpanzee Sequencing and Analysis Consortium, "Initial Sequence of the Chimpanzee Genome and Comparison with the Human Genome," *Nature* 437, no. 7055 (2005): 69–87; Aylwyn Scally, Julien Y. Dutheil, LaDeana W. Hillier,

Gregory E. Jordan, Ian Goodhead, Javier Herrero, Asger Hobolth, et al., "Insights into Hominid Evolution from the Gorilla Genome Sequence," *Nature* 483, no. 7388 (2012): 169–75.

8. Jorge J. Yunis and Ora Prakash, "The Origin of Man: A Chromosomal Pictorial Legacy," *Science* 215, no. 4539 (1982): 1525–30.

9. Yuxin Fan, Elena Linardopoulou, Cynthia Friedman, Eleanor Williams, and Barbara J. Trask, "Genomic Structure and Evolution of the Ancestral Chromosome Fusion Site in 2q13 2q14. 1 and Paralogous Regions on Other Human Chromosomes," *Genome Research* 12, no. 11 (2002): 1651 62; Mario Ventura, Claudia R. Catacchio, Saba Sajjadian, Laura Vives, Peter H. Sudmant, Tomas Marques-Bonet, Tina A. Graves, Richard K. Wilson, and Evan E. Eichler, "The Evolution of African Great Ape Subtelomeric Heterochromatin and the Fusion of Human Chromosome 2," *Genome Research* 22, no. 6 (2012): 1036–49.

10. See R. J. McKinlay Gardner, Grant R. Sutherland, and Lisa G. Shaffer, "Robertsonian Translocations," in *Chromosome Abnormalities and Genetic Counseling*, 4th ed., ed. R. J. McKinlay Gardner, Grant R. Sutherland, and Lisa G. Shaffer (Oxford: Oxford University Press, 2011).

11. See Pawel Stankiewicz, "One Pedigree We All May Have Come From—Did Adam and Eve Have the Chromosome 2 Fusion?" *Molecular Cytogenetics* 9, no. 1 (2016): 72.

12. Nick Patterson, Daniel J. Richter, Sante Gnerre, Eric S. Lander, and David Reich, "Genetic Evidence for Complex Speciation of Humans and Chimpanzees," *Nature* 441, no. 7097 (2006): 1103–108.

13. For more details on this, see Kostas Kampourakis, *Understanding Evolution* (Cambridge: Cambridge University Press, 2014), chap. 1 & 6.

CHAPTER 14: STANDING UP, WALKING UPRIGHT

1. Charles R. Darwin, *The Descent of Man, and Selection in Relation to Sex* (London: John Murray, 1871), p. 141.

2. Daniel Lieberman, *The Story of the Human Body: Evolution, Health, and Disease* (London: Penguin, 2014), pp. 34–37; William E. H. Harcourt-Smith, "The Origins of Bipedal Locomotion," in *Handbook of Paleoanthropology*, 2nd ed., ed. W. Henke and I. Tattersall (Berlin: Springer, 2015), pp. 1923–24.

3. The term "hominin" refers to any species that is more closely related to humans than chimpanzees, including modern humans themselves. Therefore, the

term "hominin" can be used to refer to all species that evolved since the last common ancestor between humans and chimpanzees, and which are more closely related to humans than to chimpanzees, including the genera *Ardipithecus*, *Australopithecus*, *Paranthropus*, and *Homo*. This is different from the term "hominid" that is used to refer to all modern and extinct great apes including gorillas, chimpanzees, orangutans, humans, and their immediate ancestors.

4. Harcourt-Smith, "Origins of Bipedal Locomotion," pp. 1939–45; William E. H. Harcourt-Smith, "Early Hominin Diversity and the Emergence of the Genus Homo," *Journal of Anthropological Sciences* 94 (2016): 19–27.

5. Ian Tattersall, *Masters of the Planet: The Search for Our Human Origins* (New York: Palgrave Macmillan, 2012), p. ix.

6. David J. Hosken and Clarissa M. House, "Sexual Selection," *Current Biology* 21, no. 2 (2011): R62–65.

7. Nancy Makepeace Tanner, *On Becoming Human* (Cambridge: Cambridge University Press, 1981), pp. 165–66.

8. Peter E. Wheeler, "The Thermoregulatory Advantages of Hominid Bipedalism in Open Equatorial Environments: The Contribution of Increased Convective Heat Loss and Cutaneous Evaporative Cooling," *Journal of Human Evolution* 21, no. 2 (1991): 107–115.

9. Graeme D. Ruxton and David M. Wilkinson, "Avoidance of Overheating and Selection for Both Hair Loss and Bipedality in Hominins," *Proceedings of the National Academy of Sciences* 108, no. 52 (2011): 20965–69.

10. Susannah K. S. Thorpe, Roger L. Holder, and Robin H. Crompton, "Origin of Human Bipedalism as an Adaptation for Locomotion on Flexible Branches," *Science* 316, no. 5829 (2007): 1328–31.

11. Craig B. Stanford, "Arboreal Bipedalism in Wild Chimpanzees: Implications for the Evolution of Hominid Posture and Locomotion," *American Journal of Physical Anthropology* 129, no. 2 (2006): 225–31.

12. Lieberman, *Story of the Human Body*, pp. 40–42.

13. Michael D. Sockol, David A. Raichlen, and Herman Pontzer, "Chimpanzee Locomotor Energetics and the Origin of Human Bipedalism," *Proceedings of the National Academy of Sciences* 104, no. 30 (2007): 12265–69.

14. David R. Carrier, "The Advantage of Standing up to Fight and the Evolution of Habitual Bipedalism in Hominins," *PLoS One* 6, no. 5 (2011): e19630.

15. Henry Gee, *The Accidental Species: Misunderstandings of Human Evolution* (Chicago: University of Chicago Press, 2013), p. 121.

16. David Pilbeam, "The Anthropoid Postcranial Axial Skeleton: Comments on

Development, Variation, and Evolution," *Journal of Experimental Zoology* 302B, no. 3 (2004): 241–67.

17. Lieberman, *Story of the Human Body*, pp. 46–47.

18. David R. Begun, *The Real Planet of the Apes: A New Story of Human Origins* (Princeton: Princeton University Press, 2016), pp. 217–20.

19. Tattersall, *Masters of the Planet*, p. 18.

CHAPTER 15: A PROLONGED BRAIN DEVELOPMENT

1. Charles R. Darwin, *The Descent of Man, and Selection in Relation to Sex* (London: John Murray, 1871), pp. 13–14.

2. Thomas Henry Huxley, *Evidence as to Man's Place in Nature* (London: Williams and Norgate, 1863), p. 67.

3. Sean B. Carroll, *Endless Forms Most Beautiful: The New Science of Evo Devo and the Making of the Animal Kingdom* (New York: W. W. Norton, 2005); Alessandro Minelli, *Forms of Becoming: The Evolutionary Biology of Development* (Princeton: Princeton University Press, 2009).

4. J. G. M. Thewissen, M. J. Cohn, L. S. Stevens, S. Bajpai, J. Heyning, and W. E. Horton, "Developmental Basis for Hind-Limb Loss in Dolphins and Origin of the Cetacean Bodyplan," *Proceedings of the National Academy of Sciences* 103, no. 22 (2006): 8414–18.

5. Karen E. Sears, Richard R. Behringer, John J. Rasweiler, and Lee A. Niswander, "Development of Bat Flight: Morphologic and Molecular Evolution of Bat Wing Digits," *Proceedings of the National Academy of Sciences* 103, no. 17 (2006): 6581–86.

6. Mary-Claire King and A. C. Wilson, "Evolution at Two Levels in Humans and Chimpanzees," *Science* 188, no. 4184 (1975), pp. 107–116.

7. Charles G. Sibley and Jon E. Ahlquist, "The Phylogeny of the Hominoid Primates, as Indicated by DNA-DNA Hybridization," *Journal of Molecular Evolution* 20, no. 1 (1984): 2–15; Ingo Ebersberger, Dirk Metzler, Carsten Schwarz, and Svante Pääbo, "Genomewide Comparison of DNA Sequences between Humans and Chimpanzees," *American Journal of Human Genetics* 70, no. 6 (2002): 1490–97.

8. Galina Glazko, Vamsi Veeramachaneni, Masatoshi Nei, and Wojciech Makałowski, "Eighty Percent of Proteins Are Different between Humans and Chimpanzees," *Gene* 346 (2005): 215–19.

9. Stephen Jay Gould, *Ontogeny and Phylogeny* (Cambridge, MA: Harvard University Press, 1977), p. 2.

10. Ibid., p. 4.

11. Wallace Arthur, *Evolution: A Developmental Approach* (Chichester: Wiley-Blackwell, 2011), pp. 93–105.

12. Gould, *Ontogeny and Phylogeny*, pp. 352–55, 386. For the respective figures, see, for instance, the photos by Adolf Naef (1926) featured in "Patterns of Neoteny in the Relative Skull Growth in *Homo* and *Pan*," 2009, https://www.mun.ca/biology/scarr/Neoteny_in_humans.htm (accessed October 17, 2017).

13. Eve K. Boyle and Bernard Wood, "Human Evolutionary History," in *Evolution of Nervous Systems* 2nd ed., vol. 4, *The Evolution of the Human Brain: Apes and Other Ancestors*, ed. T. M. Preuss (San Diego, CA: Elsevier, 2016), pp. 19–36.

14. Mehmet Somel, Henriette Franz, Zheng Yan, Anna Lorenc, Song Guo, Thomas Giger, Janet Kelso, et al., "Transcriptional Neoteny in the Human Brain," *Proceedings of the National Academy of Sciences* 106, no. 14 (2009): 5743–48.

15. Xiling Liu, Mehmet Somel, Lin Tang, Zheng Yan, Xi Jiang, Song Guo, Yuan Yuan, et al., "Extension of Cortical Synaptic Development Distinguishes Humans from Chimpanzees and Macaques," *Genome Research* 22, no. 4 (2012): 611–22.

16. Aida Gómez-Robles, William D. Hopkins, Steven J. Schapiro, and Chet C. Sherwood, "Relaxed Genetic Control of Cortical Organization in Human Brains Compared with Chimpanzees," *Proceedings of the National Academy of Sciences* 112, no. 48 (2015): 14799–804.

17. Jonas Langer, "The Heterochronic Evolution of Primate Cognitive Development," *Biological Theory* 1, no. 1 (2006): 41–43.

18. Daniel Lieberman, *The Story of the Human Body: Evolution, Health, and Disease* (London: Penguin, 2014), pp. 109–10.

19. Mehmet Somel, Xiling Liu, and Philipp Khaitovich, "Human Brain Evolution: Transcripts, Metabolites and Their Regulators," *Nature Reviews Neuroscience* 14, no. 2 (2013): 112–27.

20. Gould, *Ontogeny and Phylogeny*, p. 409.

CHAPTER 16: OUR BIOCULTURAL EVOLUTION

1. Jean-Jacques Hublin, Abdelouahed Ben-Ncer, Shara E. Bailey, Sarah E. Freidline, Simon Neubauer, Matthew M. Skinner, Inga Bergmann, et al., "New Fossils from Jebel Irhoud, Morocco and the Pan-African Origin of *Homo Sapiens*," *Nature* 546, no. 7657 (2017): 289–92; Daniel Richter, Rainer Grün, Renaud Joannes-Boyau, Teresa E. Steele, Fethi Amani, Mathieu Rué, Paul Fernandes, et al., "The Age of the

Hominin Fossils from Jebel Irhoud, Morocco, and the Origins of the Middle Stone Age," *Nature* 546, no. 7657 (2017): 293–96.

2. Matthias Krings, Anne Stone, Ralf W. Schmitz, Heike Krainitzki, Mark Stoneking, and Svante Pääbo, "Neandertal DNA Sequences and the Origin of Modern Humans," *Cell* 90, no. 1 (1997): 19–30.

3. Ralf W. Schmitz, David Serre, Georges Bonani, Susanne Feine, Felix Hillgruber, Heike Krainitzki, Svante Pääbo, and Fred H. Smith, "The Neandertal Type Site Revisited: Interdisciplinary Investigations of Skeletal Remains from the Neander Valley, Germany," *Proceedings of the National Academy of Sciences* 99, no. 20 (2002): 13342–47.

4. Richard E. Green, Johannes Krause, Adrian W. Briggs, Tomislav Maricic, Udo Stenzel, Martin Kircher, Nick Patterson, et al., "A Draft Sequence of the Neandertal Genome," *Science* 328, no. 5979 (2010): 710–22.

5. Johannes Krause, Qiaomei Fu, Jeffrey M. Good, Bence Viola, Michael V. Shunkov, Anatoli P. Derevianko, and Svante Pääbo, "The Complete Mitochondrial DNA Genome of an Unknown Hominin from Southern Siberia," *Nature* 464, no. 7290 (2010): 894–97.

6. David Reich, Richard E. Green, Martin Kircher, Johannes Krause, Nick Patterson, Eric Y. Durand, Bence Viola, et al., "Genetic History of an Archaic Hominin Group from Denisova Cave in Siberia," *Nature* 468, no. 7327 (2010): 1053–60.

7. Kay Prüfer, Fernando Racimo, Nick Patterson, Flora Jay, Sriram Sankararaman, Susanna Sawyer, Anja Heinze, et al., "The Complete Genome Sequence of a Neanderthal from the Altai Mountains," *Nature* 505, no. 7481 (2014): 43–49.

8. Svante Pääbo, "The Diverse Origins of the Human Gene Pool," *Nature Reviews Genetics* 16, no. 6 (2015): 313; see also Fernando Racimo, Sriram Sankararaman, Rasmus Nielsen, and Emilia Huerta-Sánchez, "Evidence for Archaic Adaptive Introgression in Humans," *Nature Reviews Genetics* 16, no. 6 (2015): 359–71; Fernando Racimo, Davide Marnetto, and Emilia Huerta-Sánchez, "Signatures of Archaic Adaptive Introgression in Present-Day Human Populations," *Molecular Biology and Evolution* 34, no. 2 (2017): 296–317.

9. James P. Noonan, Graham Coop, Sridhar Kudaravalli, Doug Smith, Johannes Krause, Joe Alessi, Feng Chen, et al., "Sequencing and Analysis of Neanderthal Genomic DNA," *Science* 314, no. 5802 (2006): 1113–18.

10. Richard E. Green, Johannes Krause, Susan E. Ptak, Adrian W. Briggs, Michael T. Ronan, Jan F. Simons, Lei Du, et al., "Analysis of One Million Base Pairs of Neanderthal DNA," *Nature* 444, no. 7117 (2006): 330–36;

11. Jeffrey D. Wall and Sung K. Kim, "Inconsistencies in Neanderthal Genomic DNA Sequences," *PLoS Genetics* 3, no. 10 (2007): e175.

12. John Hawks, "Significance of Neandertal and Denisovan Genomes in Human Evolution," *Annual Review of Anthropology* 42 (2013): 433–49.

13. Jonathan Marks, *Tales of the Ex-Apes: How We Think about Human Evolution* (Oakland, CA: University of California Press, 2015), pp. 165–77, p. 114.

14. Paola Villa and Wil Roebroeks, "Neandertal Demise: An Archaeological Analysis of the Modern Human Superiority Complex," *PLoS One* 9, no. 4 (2014): e96424.

15. For a book-length account that masterfully presents all of the available relevant evidence, see Joseph Henrich, *The Secret of Our Success: How Culture Is Driving Human Evolution, Domesticating Our Species, and Making Us Smarter* (Princeton: Princeton University Press, 2015).

16. Peter J. Richerson and Robert Boyd, *Not by Genes Alone: How Culture Transformed Human Evolution* (Chicago: University of Chicago Press, 2005), p. 5.

17. Luigi Luca Cavalli Sforza, *Genes, Peoples and Languages* (London: Penguin, 2001), p. 194–205.

18. Kevin N. Laland, John Odling-Smee, and Sean Myles, "How Culture Shaped the Human Genome: Bringing Genetics and the Human Sciences Together," *Nature Reviews Genetics* 11, no. 2 (2010): 137–48.

19. George H. Perry, Nathaniel J. Dominy, Katrina G. Claw, Arthur S. Lee, Heike Fiegler, Richard Redon, John Werner, et al., "Diet and the Evolution of Human Amylase Gene Copy Number Variation," *Nature Genetics* 39, no. 10 (2007): 1256–60.

20. Robert Boyd, Peter J. Richerson, and Joseph Henrich, "The Cultural Niche: Why Social Learning Is Essential for Human Adaptation," *Proceedings of the National Academy of Sciences* 108, no. supp. 2 (2011): 10918–25.

21. Henrich, *Secret of Our Success*, pp. 13–15.

22. Ibid., pp. 211–24.

23. Ibid., pp. 225–28.

CONCLUSION

1. *"Jurassic Park* (1993) Quotes," Internet Movie Database, http://www.imdb.com/title/tt0107290/quotes (accessed June 10, 2017).

2. David Jablonski, "Extinction: Past and Present," *Nature* 427, no. 6975 (2004): 589.

3. Patrick Bateson and Peter Gluckman, *Plasticity, Robustness, Development and Evolution* (Cambridge: Cambridge University Press, 2011).

4. Gregory Radick, "Is the Theory of Natural Selection Independent of Its History?" in *The Cambridge Companion to Darwin*, ed. J. Hodge and G. Radick (Cambridge: Cambridge University Press, 2003), pp. 143–67.

5. Simon Conway Morris, *Life's Solution: Inevitable Humans in a Lonely Universe* (Cambridge: Cambridge University Press, 2003).

6. Jonathan Losos, *Improbable Destinies: Fate, Chance and the Future of Evolution* (New York: Riverhead, 2017), p. 315.

7. Walter Alvarez, *A Most Improbable Journey: A Big History of Our Planet and Ourselves* (W. W. Norton, 2016), p. 184.

8. Ibid., pp. 184–87.

9. Michael J. Harms and Joseph W. Thornton, "Historical Contingency and Its Biophysical Basis in Glucocorticoid Receptor Evolution," *Nature* 512, no. 7513 (2014): 203.

10. Tyler N. Starr, Lora K. Picton, and Joseph W. Thornton, "Alternative Evolutionary Histories in the Sequence Space of an Ancient Protein," *Nature* 549, no. 7672 (2017): 409–13.

11. Stephen Jay Gould, *The Structure of Evolutionary Theory* (Cambridge, MA: Harvard University Press, 2002), p. 1333.

12. Carl Hoefer, "Causal Determinism," in *The Stanford Encyclopedia of Philosophy*, spring 2016 ed., edited by Edward N. Zalta (Stanford, CA: Stanford University, 2016), https://plato.stanford.edu/archives/spr2016/entries/determinism -causal/ (accessed October 18, 2017); first published Jan. 23, 2003, substantively revised Jan. 21, 2016.

13. Baron Reed, "Certainty," in *The Stanford Encyclopedia of Philosophy*, winter 2011 ed., edited by Edward N. Zalta (Stanford, CA: Stanford University, 2011), https://plato.stanford.edu/archives/win2011/entries/certainty/ (accessed September 13, 2017).

14. Carl G. Hempel and Paul Oppenheim, "Studies in the Logic of Explanation," *Philosophy of Science* 15, no. 2 (1948): 135–75.

15. David L. Hull, "The Particular-Circumstance Model of Scientific Explanation," in *History and Evolution*, ed. M. H. Nitecki and D. V. Nitecki (Albany: State University of New York Press, 1992), pp. 69–80. The deductive-nomological model of explanation is not only outdated but also problematic; however, I am using it here because it makes the contrast between physical and biological explanations clear.

16. Gould, *Structure of Evolutionary Theory*, pp. 1333–34.

17. Stephen Jay Gould, *Wonderful Life: The Burgess Shale and the Nature of History* (1989; Vintage: London, 2000), p. 291.

18. Simon Conway Morris, *The Crucible of Creation: The Burgess Shale and the Rise of Animals* (Oxford: Oxford University Press, 1998), p. 14.

19. This statement comes from Kazantzakis's book *Ασκητική*. "Ξέρω τώρα˙ δεν ελπίζω τίποτα, δε φοβούμαι τίποτα, λυτρώθηκα από το νου κι από την καρδιά, ανέβηκα πιο πάνω, είμαι λεύτερος. Αυτό θέλω. Δε θέλω τίποτα άλλο. Ζητούσα ελευτεριά." Νίκος Καζαντζάκης, *Ασκητική* (Αθήνα: Εκδόσεις Ελ. Καζαντζάκη, 1971), p.25. This means "I know now; I hope for nothing. I fear nothing, I have liberated myself from both the mind and the heart, I have gone up, I am free. This is what I want. I want nothing more. I was seeking freedom." (This book has been translated to English under the title *The Saviors of God: Spiritual Exercises*.)

INDEX

on the difficulties of his theory, 222–23

early thoughts on transmutation, 210

on first cause, 245

on his grandfather's *Zoonomia*, 209

and Jean-Baptiste Lamarck, 198, 223–24, 226

life before the *Beagle* voyage, 198–200

and natural selection, 214–19, 236–37

and the principle of divergence, 228–32, 237

on similarities between humans and apes, 246, 279–80

and Thomas Malthus, 213–17

and variation, 228–29

and the *Vestiges of the Natural History of Creation*, 225–28

on William Paley, 203

Darwin, Emma, 225

Darwin, Erasmus (Charles's brother), 198, 218

Darwin, Erasmus (Charles's grandfather), 198, 209

Darwin, Robert Waring, 198, 201

Dembski, William, 108

Denisovans, 293–95, 297–301

design, 21, 46

  argument from, 105

  definition of, 18, 38

  and blind watchmaker metaphor (Dawkins), 105

and genes as blueprints, 81

intelligent design, 19, 35, 37, 38, 40, 50, 64, 104, 111–120, 245, 298, 310

and Plato, 44–45

and purpose, 46, 49, 51–52, 58, 60–62, 64, 96, 101

and significant events, 85

and teleology, 63–64

and watchmaker metaphor (Paley), 108–109

without designer (natural), 63

See also designed/non-designed distinction; design stance

designed/non-designed distinction, 63–64

design stance

  consequences of, 64–65, 120, 310

  definition of, 57–59

  and genes, 81, 83

  how to overcome, 321, 325

  and intelligent design, 105

  and purpose in life events, 98, 100

destiny, 21, 23, 35, 36, 38, 40, 64, 79, 120, 310, 321

  and astrology, 90, 92

  definition of, 18–19

  and reincarnation, 99–100

  and teleology, 94

determinism

  causal, 321–24

  definition of, 17, 67

  genetic, 67–68